REVISE EDEXCEL GCSE (9–1)

Physics

REVISION GUIDE

Higher

Series Consultant: Harry Smith
Author: Dr Mike O'Neill

A note from the publisher

In order to ensure that this resource offers high-quality support for the associated Pearson qualification, it has been through a review process by the awarding body. This process confirms that this resource fully covers the teaching and learning content of the specification or part of a specification at which it is aimed. It also confirms that it demonstrates an appropriate balance between the development of subject skills, knowledge and understanding, in addition to preparation for assessment.

Endorsement does not cover any guidance on assessment activities or processes (e.g. practice questions or advice on how to answer assessment questions), included in the resource nor does it prescribe any particular approach to the teaching or delivery of a related course.

While the publishers have made every attempt to ensure that advice on the qualification and its assessment is accurate, the official specification and associated assessment guidance materials are the only authoritative source of information and should always be referred to for definitive guidance.

Pearson examiners have not contributed to any sections in this resource relevant to examination papers for which they have responsibility.

Examiners will not use endorsed resources as a source of material for any assessment set by Pearson.

Endorsement of a resource does not mean that the resource is required to achieve this Pearson qualification, nor does it mean that it is the only suitable material available to support the qualification, and any resource lists produced by the awarding body shall include this and other appropriate resources.

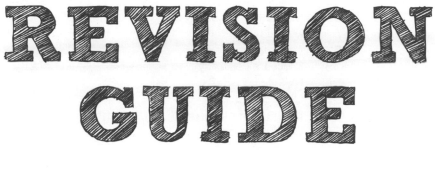

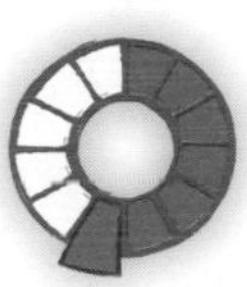

Question difficulty

Look at this scale next to each exam-style question. It tells you how difficult the question is.

Contents

1-to-1 page match with the Physics Higher Revision Workbook ISBN 9781292133683

A small bit of small print:
Edexcel publishes Sample Assessment Material and the Specification on its website. This is the official content and this book should be used in conjunction with it. The questions in Now try this have been written to help you practise every topic in the book. Remember: the real exam questions may not look like this.

Key concepts

You need to know the SI units of physical quantities, as well as their multiples and sub-multiples, how to convert between units and how to use significant figures and standard form.

Base SI units

There are six main **base SI units** you need to know.

Name	Unit	Abbreviation
length	metre	m
mass	kilogram	kg
time	second	s
current	ampere or amp	A
temperature	kelvin	K
amount of a substance	mole	mol

Derived SI units

There are nine derived SI units you need to know.

Name	Unit	Abbreviation
frequency	hertz	Hz
force	newton	N
energy	joule	J
power	watt	W
pressure	pascal	Pa
electric charge	coulomb	C
electric potential difference	volt	V
electric resistance	ohm	Ω
magnetic flux density	tesla	T

Multiples and sub-multiples

For values of physical quantities, you need to know the values of these multiples or **prefixes**.

10^9 10^8 10^7 10^6 10^5 10^4 10^3 10^2 10^1 10^0 10^{-1} 10^{-2} 10^{-3} 10^{-4} 10^{-5} 10^{-6} 10^{-7} 10^{-8} 10^{-9}

giga-	mega-	kilo-	centi-	milli-	micro-	nano-
prefix **G**	prefix **M**	prefix **k**	prefix **c**	prefix **m**	prefix **μ**	prefix **n**
× 1 000 000 000	× 1 000 000	× 1000	÷ 100	÷ 1000	÷ 1 000 000	÷ 1 000 000 000
e.g. 5 GHz	e.g. 5 MW	e.g. 2 km	e.g. 8 cm	e.g. 6 mA	e.g. 7 μs	e.g. 2 nm

Significant figures

Significant figures are either non-zero digits (i.e. the digits 1 to 9) or zeroes between non-zero digits.

- The number **543 563** has 6 significant figures.
- The number 0.**23** has two significant figures.
- The number 0.000 000 000 **254** has 3 significant figures. The zeroes before the '2' are not significant figures.
- The number **3034**0000 has 4 significant figures.

The number 30 339 900 rounded to 5 significant figures is 30 340 000.

Standard form

It is useful to write very large and very small numbers in **standard form**.

$$A \times 10^B$$

A: This is a number between 1 and 10

B: This is an integer – a whole number that can be positive or negative.

You also need to be able to convert between units such as kilometres to metres, or hours to seconds.

Now try this

1 State the number of significant figures in each number.
(a) 4.56 **(1 mark)**
(b) 0.000 564 5 **(1 mark)**
(c) 3.0046 **(1 mark)**

2 Convert:
(a) 12 hours to seconds **(2 marks)**
(b) 64 km/h to m/s **(2 marks)**

3 Convert 58.3 MW to watts. Give your answer in standard form. **(2 marks)**

4 The speed of light is 186 000 miles per second. Convert this to metres per second and write it in standard form. **(3 marks)**

1 mile is equal to 1609 m

Scalars and vectors

All physical quantities can be described as either a scalar or a vector quantity.

Scalar quantities

Scalar quantities have a size or a magnitude but no specific direction.

Examples include:

- mass
- speed
- distance
- energy
- temperature.

Vector quantities

Vector quantities have a size or magnitude and a specific direction.

Examples include:

- force or weight
- velocity
- displacement
- acceleration
- momentum.

Speed and velocity

The girl is running to the right so she has a velocity of 5 m/s to the right.

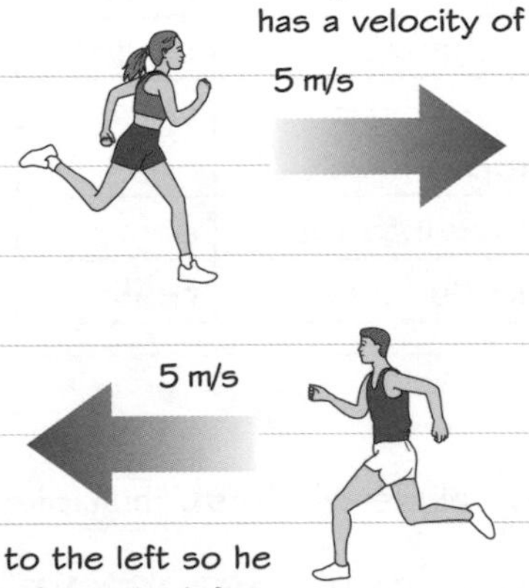

Both the girl and the boy are running at 5 m/s. They have the same speed.

If we take 'to the right' as the positive direction, then the girl has a velocity of +5 m/s and the boy has a velocity of −5 m/s.

The boy is running to the left so he has a velocity of 5 m/s to the left.

Speed has a size but velocity has a size AND a direction.

Velocity is speed in a stated direction.

Worked example

(a) Describe the difference between a scalar and a vector. **(2 marks)**

A scalar has a size, whereas a vector has a size and a specific direction.

(b) Describe an example which shows the difference between a scalar and a vector. **(2 marks)**

Speed is a scalar quantity and may have a value of 5 m/s. Velocity may also have a value of 5 m/s but a direction of north.

All vector quantities can be given positive and negative values to show their direction. Examples include:

1. A force of +4 N may be balanced by a force acting in the opposite direction of −4 N.
2. A car that accelerates at a rate of $+2\,m/s^2$ could decelerate at $-2\,m/s^2$.
3. If a distance walked to the right of 20 m is a displacement of +20 m, then the same distance walked to the left from the same starting point is a displacement of −20 m.

Now try this

1 (a) Which of these quantities is a scalar? **(1 mark)**

☐ **A** velocity ☐ **B** acceleration ☐ **C** mass ☐ **D** weight

(b) Which of these quantities is a vector? **(1 mark)**

☐ **A** temperature ☐ **B** energy ☐ **C** speed ☐ **D** electric field

2 A boy has a speed of 4 m/s when running. State his (a) speed and (b) velocity when he is running in the opposite direction. **(2 marks)**

3 Explain why a satellite can be said to be moving at a constant speed but not at a constant velocity. **(3 marks)**

Speed, distance and time

Speed is the distance that a moving object covers each second.

Calculating speed

When a body covers the same distance per second throughout its journey, you can use this equation to calculate its speed:

speed (m/s) = change in distance (m) ÷ time taken (s)

The greater the change in distance per second, the faster the object is moving.

When the change in distance over a period of time is zero, the speed is zero and the object is stationary.

Average speed

LEARN IT! IT'S NOT ON THE EQUATIONS LIST

Objects often change speed during a journey, so it is better to use average speed:

average speed (m/s) = total distance travelled (m) ÷ total time taken (s)

You can also write this as

distance travelled (m) = average speed (m/s) × time taken (s)

Distance/time graphs

Distance/time graphs have distance on the *y*-axis and time on the *x*-axis. The gradient or slope of the graph tells us about the motion of the vehicle.

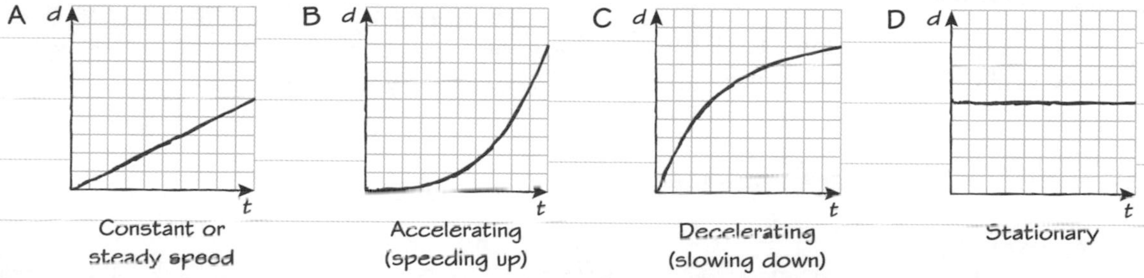

Journeys and distance/time graphs

This distance/time graph tells you about a student's journey.

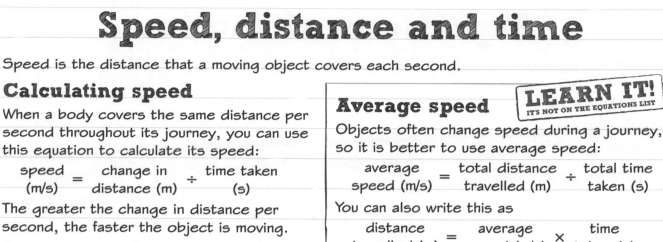

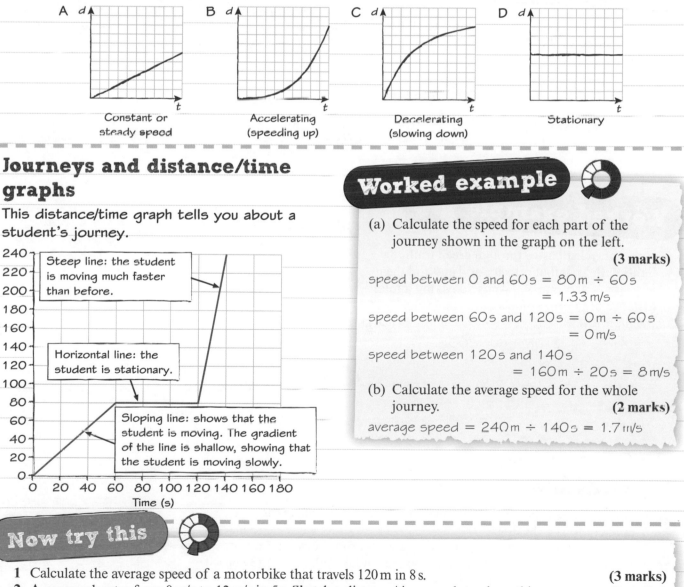

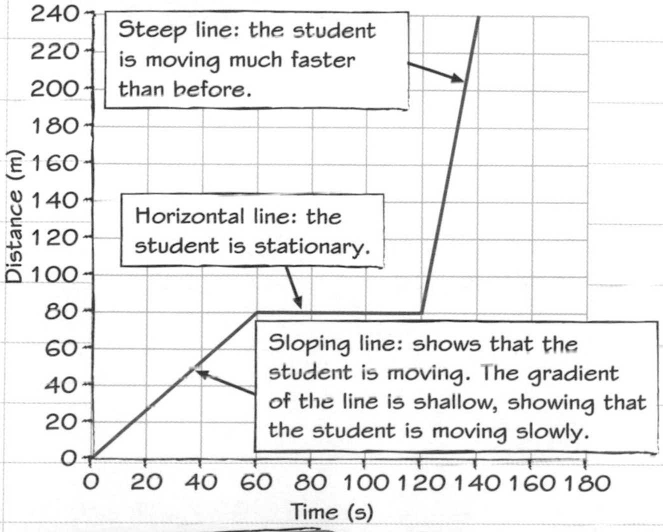

Worked example

(a) Calculate the speed for each part of the journey shown in the graph on the left. **(3 marks)**

speed between 0 and 60 s = 80 m ÷ 60 s
= 1.33 m/s

speed between 60 s and 120 s = 0 m ÷ 60 s
= 0 m/s

speed between 120 s and 140 s
= 160 m ÷ 20 s = 8 m/s

(b) Calculate the average speed for the whole journey. **(2 marks)**

average speed = 240 m ÷ 140 s = 1.7 m/s

Now try this

1 Calculate the average speed of a motorbike that travels 120 m in 8 s. **(3 marks)**
2 A car accelerates from 0 m/s to 12 m/s in 5 s. Sketch a distance/time graph to show this. **(3 marks)**
3 The speed limit for a car on a motorway in the UK is 112.7 km/h. Calculate this speed in m/s. **(3 marks)**

Equations of motion

You can use equations to work out the velocity and acceleration of moving bodies.

Acceleration

Acceleration is a change in velocity per second. Acceleration is a vector quantity.

$$\text{acceleration (m/s}^2) = \frac{\text{change in velocity (m/s)}}{\text{time taken (s)}}$$

$$a = \frac{(v - u)}{t}$$

LEARN IT! IT'S NOT ON THE EQUATIONS LIST

a is the acceleration
v is the final velocity
u is the initial velocity
t is the time taken

Velocity

Velocity is the change in distance per second.

Velocity is a vector quantity.

(final velocity)^2 – (initial velocity)^2 = 2 × acceleration × distance

$$v^2 - u^2 = 2 \times a \times x$$

x is the distance travelled.

You can also write this as:

$$v^2 = u^2 + 2ax$$

Worked example

A cat changes its speed from 2.5 m/s to 10.0 m/s over a period of 3.0 s. Calculate the cat's acceleration. **(3 marks)**

v = 10 m/s, u = 2.5 m/s, t = 3 s, a = ?
$a = (v - u) \div t$
a = (10.0 m/s – 2.5 m/s) ÷ 3.0 s
a = 7.5 m/s ÷ 3.0 s
a = 2.5 m/s^2

Write down all the quantities you know and the one you need to work out. Then decide which equation you need to use.

Watch out! When using this equation, the value for $(v - u)$ may be negative, which means that the cat is slowing down.

Worked example

A motorcyclist passes through green traffic lights with an initial velocity of 4 m/s and then accelerates at a rate of 2.4 m/s^2, covering a total distance of 200 m. Calculate the final velocity of the motorcycle. **(4 marks)**

u = 4 m/s, a = 2.4 m/s^2, x = 200 m, v = ?
$v^2 = u^2 + 2ax$
$v^2 = (4)^2 + 2 \times 2.4 \times 200$
$v^2 = 16 + 960 = 976$
$v = \sqrt{976}$ = 31.2 m/s

Maths skills When working out x using this equation, you will need to rearrange it. Subtract u^2 from both sides and then divide both sides by $2a$, to give $x = \frac{v^2 - u^2}{2a}$. Substitute the other values and work out the value of x.

Watch out! When you substitute the values for u, a and x, you get a value for v^2. So you need to find the square root of this value to get the value for the final velocity, v.

Now try this

1 A dog changes its speed from 2 m/s to 8 m/s in 5 s. Calculate the acceleration of the dog. **(3 marks)**

2 An aeroplane starts at rest and accelerates at 1.6 m/s^2 down a runway. After 1.8 km it takes off. Calculate its speed at take-off. **(3 marks)**

3 A car passes through traffic lights at a speed of 5 m/s and then accelerates at 1.2 m/s^2 until it has reached a final speed of 18 m/s. Calculate the distance it has travelled from the traffic lights. **(4 marks)**

If a vehicle is said to be 'starting from rest' then its initial velocity, u, will be zero, so the equation simplifies to $v^2 = 2ax$.

Velocity/time graphs

Velocity/time graphs show how the velocity of a vehicle changes with time. You can also work out acceleration and distance travelled from the graph.

Interpreting velocity/time graphs

Velocity/time graphs have velocity plotted on the *y*-axis and time plotted on the *x*-axis. The graph shows you how the velocity changes with time.

- The **slope** or **gradient** of the graph tells us the acceleration of the vehicle.
- The **area under the graph** tells us the **distance** travelled.

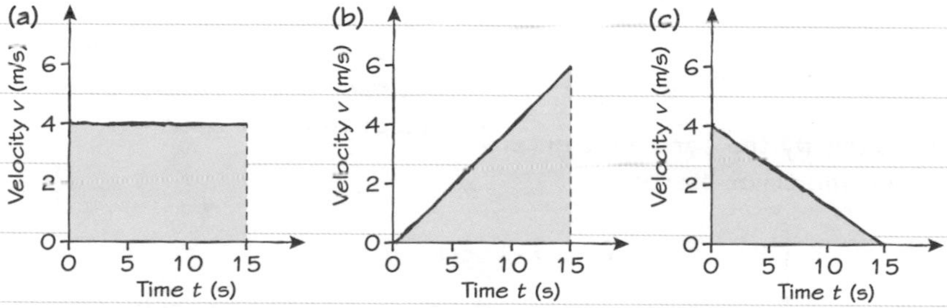
(a)
(b)
(c)
Velocity v (m/s)
0
2
4
6
0
5
10
15
Time t (s)

Maths skills Acceleration (gradient) = change in velocity/change in time. For distance travelled (area) use the formulae for the area of a rectangle (base × height, $b \times h$) and area of a triangle ($\frac{1}{2} \times b \times h$). The area will have units of metres, as m/s × s = m.

(a) Acceleration is zero, distance travelled is 60 m.

(b) Acceleration is 0.4 m/s^2, distance travelled is 45 m.

(c) Acceleration is –0.27 m/s^2, distance travelled is 30 m.

Velocity/time graphs

This velocity/time graph shows how the velocity of a train along a straight track changes with time.

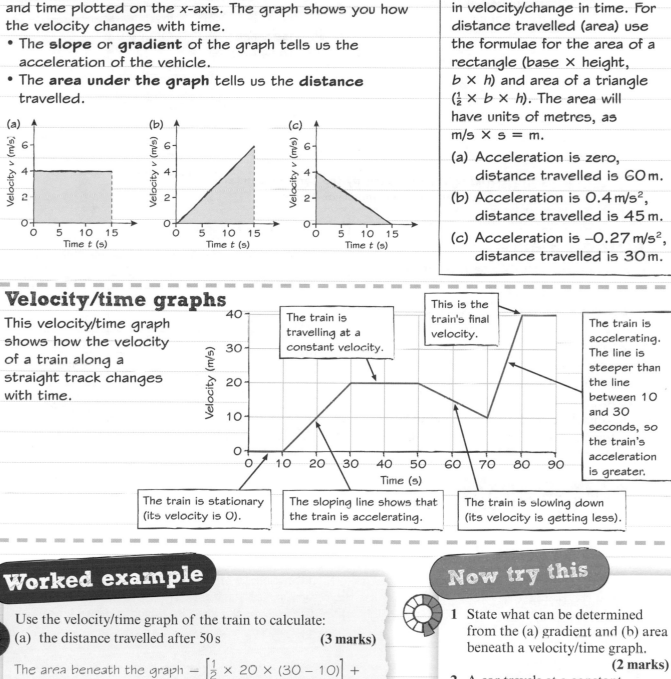

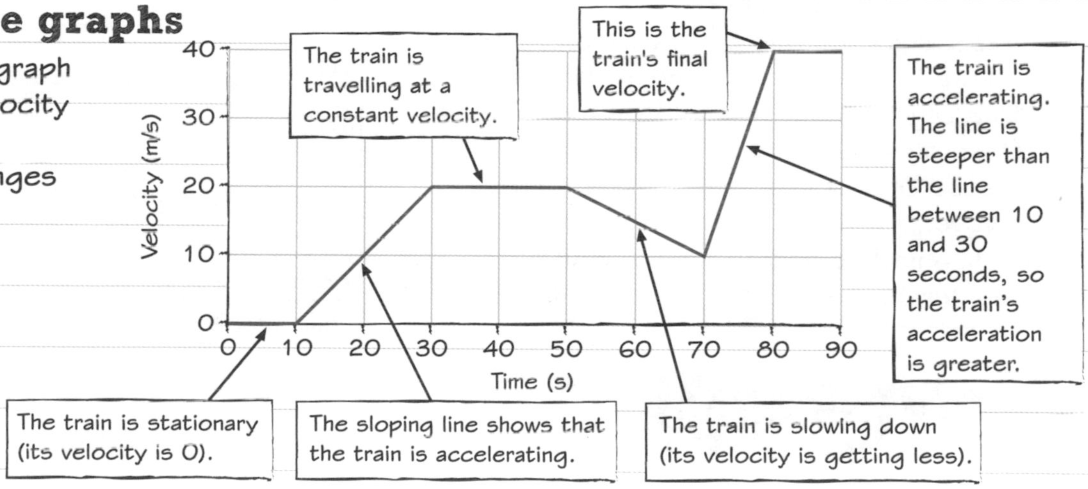
Velocity (m/s)
0
10
20
30
40
Time (s)
0
10
20
30
40
50
60
70
80
90
The train is travelling at a constant velocity.
This is the train's final velocity.
The train is accelerating. The line is steeper than the line between 10 and 30 seconds, so the train's acceleration is greater.
The train is stationary (its velocity is 0).
The sloping line shows that the train is accelerating.
The train is slowing down (its velocity is getting less).

Worked example

Use the velocity/time graph of the train to calculate:

(a) the distance travelled after 50 s **(3 marks)**

The area beneath the graph = $\left[\frac{1}{2} \times 20 \times (30 - 10)\right] + [20 \times (50 - 30)]$
= 200 m + 400 m = 600 m

(b) the acceleration of the train between 50 s and 70 s. **(3 marks)**

Change in velocity = 10 m/s – 20 m/s = –10 m/s
Change in time = 70 s – 50 s = 20 s
Acceleration = –10 m/s ÷ 20 s = –0.5 m/s^2

Now try this

1 State what can be determined from the (a) gradient and (b) area beneath a velocity/time graph. **(2 marks)**

2 A car travels at a constant speed of 8 m/s for 12 s before accelerating at 1.5 m/s^2 for the next 6 s.

(a) Sketch a velocity/time graph for the car. **(4 marks)**

(b) Calculate the total distance travelled by the car. **(4 marks)**

Determining speed

Speed can be determined in the laboratory using light gates and other equipment.

You can measure speeds of objects in the lab by using light gates connected to a computer or data logger.

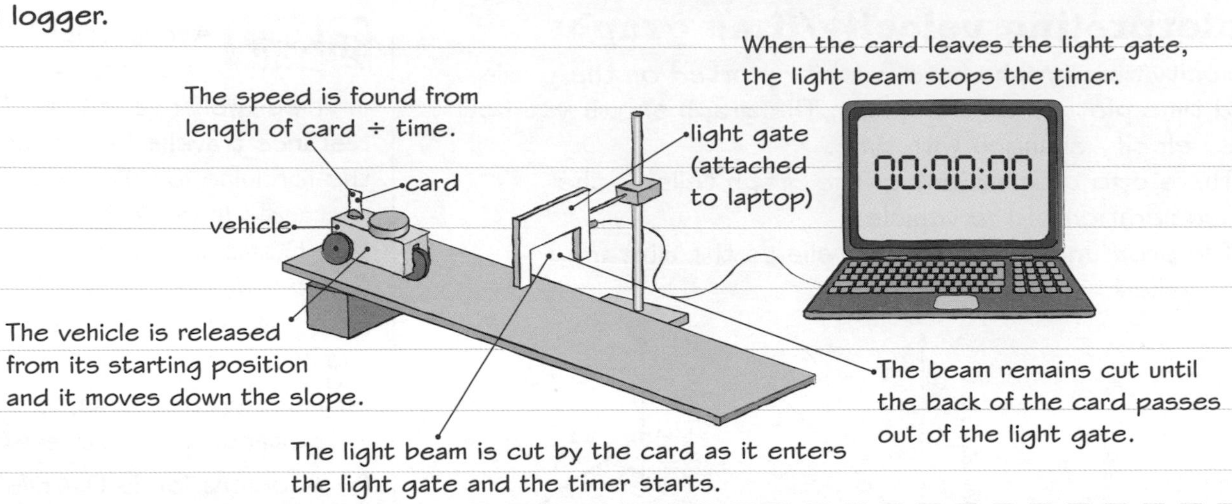

When the card leaves the light gate, the light beam stops the timer.
The speed is found from length of card ÷ time.
card
vehicle
light gate (attached to laptop)
00:00:00
The vehicle is released from its starting position and it moves down the slope.
The beam remains cut until the back of the card passes out of the light gate.
The light beam is cut by the card as it enters the light gate and the timer starts.

Falling objects

You can measure the speed of falling objects using light gates.

The acceleration due to gravity on Earth is close to 10 m/s². This means that a falling object will increase its speed by 10 m/s every second when falling in the absence of frictional or resistive forces. However, for a piece of falling paper this terminal velocity value will decrease considerably because of resistive forces.

Typical speeds

Different activities occur at different speeds.

Activity	Typical speed
walking	1.5 m/s
running	3.0 m/s
cycling	6.0 m/s
driving	14 m/s
speed of sound in air	330 m/s
airliner	250 m/s
commuter train	55 m/s
gale-force wind	16 m/s

Worked example

Describe how you can determine the average speed of a moving trolley as it moves down a slope. **(3 marks)**

When you know the distance that a body moves, and the time taken for the body to move that distance, you can use the equation speed = distance ÷ time to determine its speed. This can be done by measuring the distance between two points that the trolley moves past and dividing it by the time taken to travel between those points, or by using light gates to measure the time for which a known length of card breaks a light beam.

Now try this

1 State what you need to measure to determine the speed of an object. **(2 marks)**

2 Explain how two light gates can be used to find the acceleration of a ball rolling down a slope. **(4 marks)**

3 Describe how a falling ball can be used to find an accurate value for the acceleration due to gravity, *g*. **(5 marks)**

Newton's first law

A body will remain at rest or continue in a straight line at a constant speed as long as the forces acting on it are balanced.

Stationary bodies

The forces acting on a stationary body are balanced.

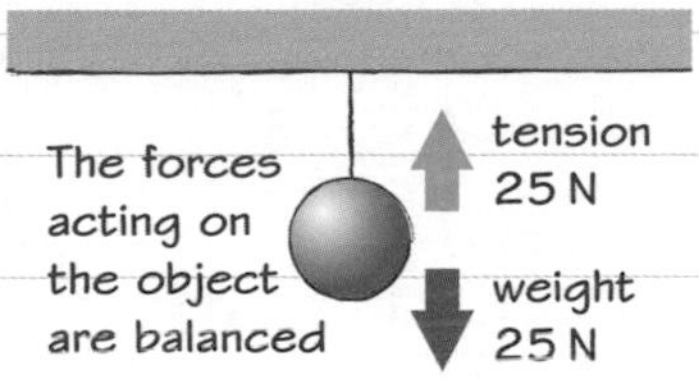

A common mistake is to think that when the resultant force on an object is zero, the object is stationary. The object may also be travelling at a constant speed.

Bodies moving at a constant speed

The forces acting on a body moving at a constant speed, and in a straight line, are balanced.

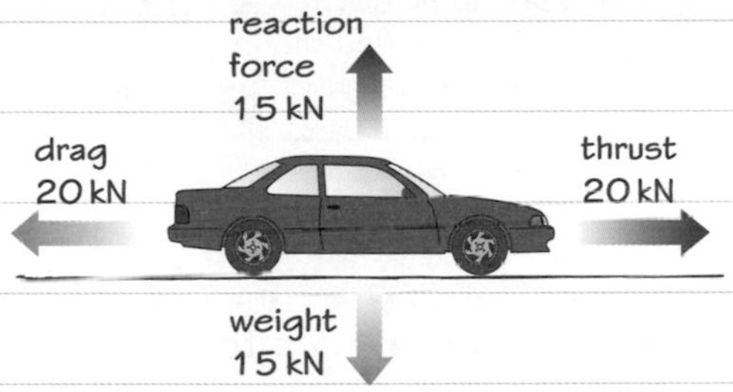

The forces on the car are balanced. The car will continue to move at a constant speed in a straight line until another external force is applied.

Unbalanced forces

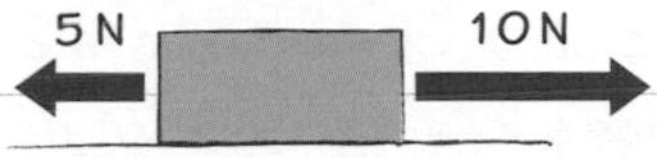

This body will accelerate to the right, since there is a resultant force of 5 N acting to the right.

Worked example

Explain the effect that each of these forces will have on a car.

(a) 300 N forward force from the engine, 200 N drag. **(3 marks)**

Resultant force = 300 N − 200 N = 100 N. The car will accelerate in the direction of the resultant force. Its velocity will increase.

(b) 200 N forward force from the engine, 400 N friction from brakes. **(3 marks)**

Resultant force = 200 N − 400 N = −200 N (200 N acting backwards). The car will accelerate in the direction of the resultant force. This is in the opposite direction to its velocity, so the car will slow down.

(c) 300 N forward force, 300 N drag. **(3 marks)**

Resultant force = 0 N. The car will continue to move at the same velocity.

A resultant force acting in the opposite direction to the movement of a body will slow it down. It can also reverse the direction of motion.

Now try this

1 State what forces act on a body that is moving at a constant speed. **(2 marks)**

2 Explain how the direction of a moving body can be made to change. **(2 marks)**

3 Draw diagrams to show how two forces of 100 N acting on a mass can make it:

(a) stationary **(2 marks)**

(b) move at a constant speed **(2 marks)**

(c) accelerate to the left. **(2 marks)**

Newton's second law

When a resultant force acts on a mass then there will be a change in its velocity. The resultant force determines the size and direction of the subsequent acceleration of the mass.

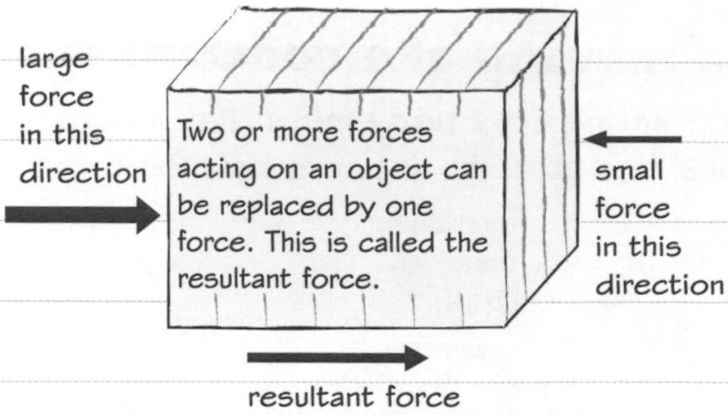

Forces have direction, so a force of −1 N is in the opposite direction to a force of 1 N.

When two or more forces act on the same straight line or are parallel, they can be added together to find the **resultant force**.

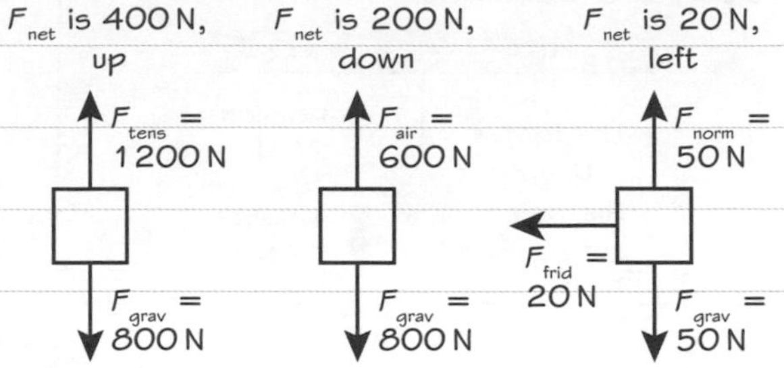

Force, mass and acceleration

You can calculate the acceleration of an object when you know its mass and the resultant force acting on it using the equation:

$$\text{acceleration (m/s}^2\text{)} = \frac{\text{force (N)}}{\text{mass (kg)}}$$

$$a = \frac{F}{m}$$

Rearrange the equation to give the force:

$$F = m \times a$$

LEARN IT! NOT ON THE EQUATIONS LIST

- The acceleration is in the same direction as the force.
- When the resultant force is zero, the acceleration is zero.
- A negative force means that the object is accelerating backwards or is slowing down.

Inertial mass is a measure of how difficult it is to change the velocity of a moving object and is defined as the ratio 'force over acceleration'.

Worked example

The diagram shows the horizontal forces acting on a boat. The boat has a mass of 400 kg.

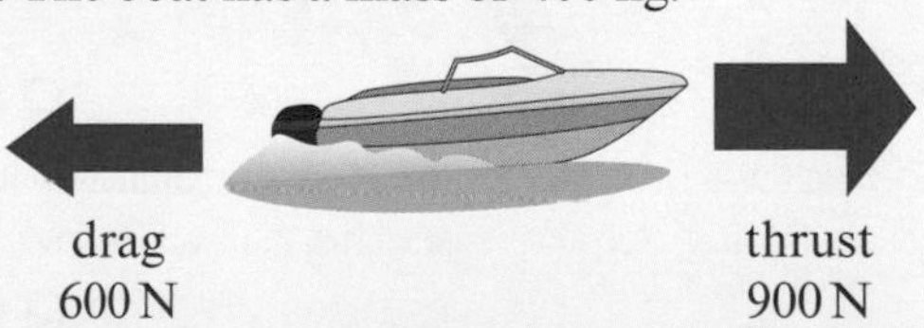

Calculate the acceleration of the boat at the instant shown in the diagram. **(3 marks)**

resultant force on boat = 900 N − 600 N = 300 N forwards

mass = 400 kg

acceleration $= \frac{F}{m} = \frac{300 \text{ N}}{400 \text{ kg}}$

$= 0.75 \text{ m/s}^2$.

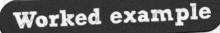

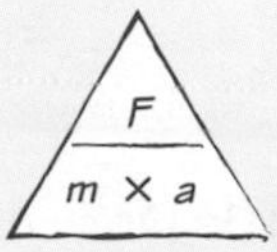

Worked example

A basketball player catches a ball. The force acting on the ball is −1.44 N and its acceleration is -2.4 m/s^2.

(a) Calculate the mass of the ball. **(3 marks)**

$m = \frac{F}{a}$

$= \frac{-1.44 \text{ N}}{-2.4 \text{ m/s}^2} = 0.6 \text{ kg}$

(b) Which of the following describes the effect of the force on the ball? **(1 mark)**

☐ **A** The ball is moving backwards.

☒ **B** The ball is slowing down moving forwards.

☐ **C** The ball is moving faster forwards.

☐ **D** The ball is slowing down moving backwards.

Make sure you include minus signs in your calculations where necessary.

Now try this

1 Calculate the resultant force that causes a 1.2 kg mass to accelerate at 8 m/s^2. **(2 marks)**

2 Calculate the size of the mass that accelerates at 0.8 m/s^2 when the resultant force is 18.8 N. **(3 marks)**

3 Calculate the acceleration of a mass of 80 g when the resultant force is 0.6 kN. **(5 marks)**

Weight and mass

It is important that you understand the difference between weight and mass. These words are often used interchangeably but are actually different things.

Weight

Weight is the **force** that a body experiences due to its mass and the size of the gravitational field that it is in.

Weight is a **vector** quantity and is measured in **newtons** (N).

The weight of a body on the surface of the Earth acts inwards towards the Earth's centre.

Mass

Mass is a measure of the **amount of matter** that is contained within a three-dimensional space.

Mass is a **scalar** quantity and is measured in **kilograms** (kg).

The mass of a body is not affected by the size of the gravitational field that it is in.

Connection between mass and weight

To find the weight of an object, use the equation:

weight (N) = mass (kg) × gravitational field strength (N/kg)

$W = m \times g$

LEARN IT! IT'S NOT ON THE EQUATIONS LIST

The weight of an object is directly proportional to the value of g, so a mass will weigh more on Earth than it does on the Moon.

Worked example

An astronaut has a mass of 58 kg on Earth. State the astronaut's mass and weight:

(a) on the surface of the Moon **(3 marks)**

Mass does not change so it is still 58 kg.

weight = 58 kg × 1.6 N/kg = 92.8 N

(b) on the surface of Jupiter. **(3 marks)**

Mass does not change, so it is still 58 kg.

weight = 58 kg × 26 N/kg = 1508 N

Gravitational field strength

The gravitational field strength of a body, such as a planet, depends on:

1. the **mass** of the body
2. the **radius** of the body.

When a body has a large mass and a small radius, it will have a large gravitational field strength.

The units of gravitational field strength are newtons per kilogram (N/kg), but it can also be given as m/s^2.

Astronomical body	Value for g
Earth	10 N/kg
Moon	1.6 N/kg
Jupiter	26 N/kg
Neptune	13.3 N/kg
Mercury	3.6 N/kg
Mars	3.75 N/kg
neutron star	10^{12} N/kg

Measuring weight

Weight is measured using a newtonmeter.

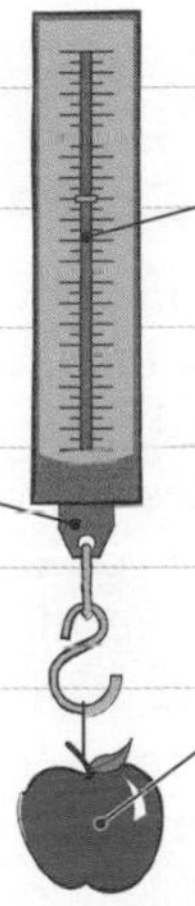

The greater the mass attached, the more weight it will experience due to gravity and the more the spring will stretch.
Reading the scale tells you the weight of the mass in newtons.
A 2 kg mass has twice the weight on Earth as a 1 kg mass, so the extension of the spring will be twice as big.

Now try this

1 Calculate the value for g on the surface of a planet where a mass of 18 kg experiences a weight of 54 N. **(3 marks)**

2 The value of g on Earth is 10 N/kg. On Planet X, the value of g is 25 N/kg. Explain what the mass and weight of a 5 kg brick will be on the surface of Planet X. **(3 marks)**

Had a look ☐ Nearly there ☐ Nailed it! ☐

Force and acceleration

Practical skills You can determine the relationship between force, mass and acceleration by varying the masses added to a trolley and measuring the time it takes to pass between two light gates that are a small distance apart.

Core practical

There is more information about the relationship between force, mass and acceleration on pages 8 and 9.

Investigating force and acceleration

Aim

To investigate the effect of mass on the acceleration of a trolley.

Apparatus

trolley, light gates, data logger, card of known length, slope or ramp, masses

An accelerating mass of greater than a few hundred grams can be dangerous, and may hurt somebody if it hits them at speed. Bear this in mind when designing your investigation.

Method

1 Set up the apparatus as shown.

It is better to use light gates and other electronic equipment to record values as this is more accurate than using a ruler and a stopwatch.

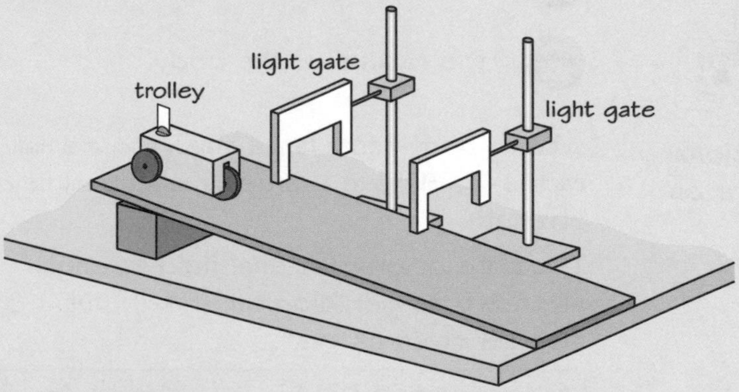

2 Set up the light gates to take the velocity and time readings for you.

3 Record velocity and time for different values of mass on the trolley.

4 Work out acceleration by dividing the difference in the velocity values by the time taken for the card to pass between both gates.

5 If you are changing the mass, the slope or gradient needs to remain the same throughout the investigation.

Results

The results for a slope of 20° to the horizontal and a card width of 10 cm are shown in the table below.

Mass (g)	Δv (m/s)	Δt (s)	a (m/s^2)
200	7.0	2.2	3.2
400	6.7	2.1	3.2
600	6.8	2.2	3.1
800	6.9	2.1	3.3

Conclusion

The acceleration of the trolley does not depend on the mass of the trolley and will remain fairly constant throughout.

Maths skills: Velocity and acceleration

Key points to remember for this investigation are:

Acceleration is change in speed ÷ time taken, so two velocity values are needed, along with the time difference between these readings, to obtain a value for the acceleration of the trolley.

Velocity is the rate of change of displacement and acceleration is the rate of change of velocity. The word rate means 'per unit time':

$v = \Delta x \div \Delta t$ and $a = \Delta v \div \Delta t$

Now try this

1 Explain why it is better to use light gates and a data logger than a stopwatch to record time values. **(3 marks)**

2 Suggest how the investigation described above could be improved to obtain better results. **(4 marks)**

3 Design a similar experiment to determine how acceleration depends on the gradient of the slope. **(5 marks)**

Circular motion

A body will move in a circular path if its motion is at right angles to a force that acts inwards along the radius of the path.

Speed and velocity

A body moving in a circular path with **constant speed** still has a **changing velocity**.

Speed is a **scalar** quantity and has size but no direction. Velocity is a **vector** quantity, which means it has both size and direction.

The velocity of a body changes when **either** its **speed** or its **direction** of motion changes.

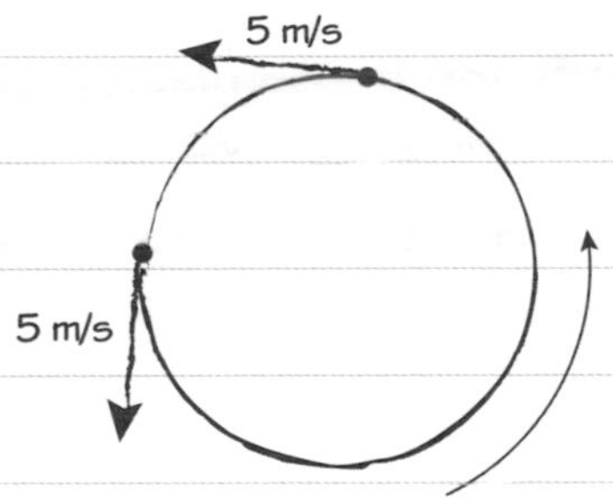

Why circular motion happens

For circular motion to occur there must be a force acting inwards along the radius of the circle. This is called the **centripetal force**. The object moves at **right angles** to this force.

One example of this is the motion of a satellite in orbit around the Earth.

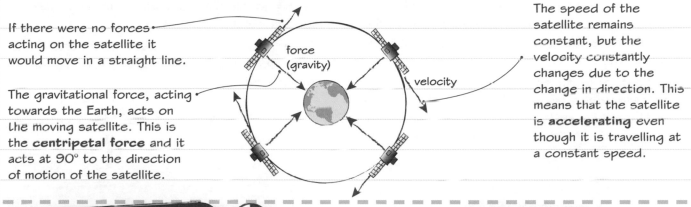

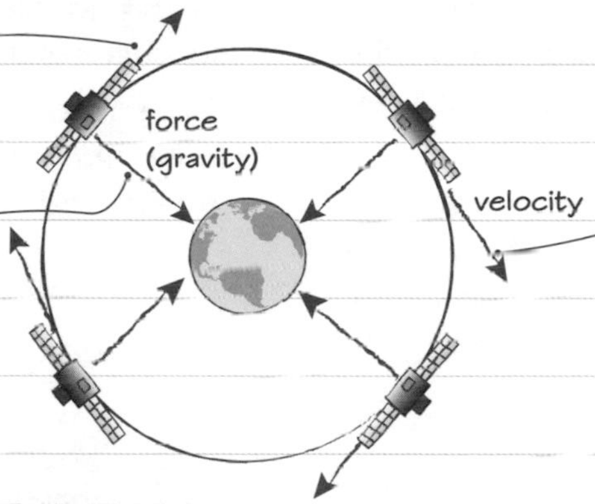

If there were no forces acting on the satellite it would move in a straight line.
The gravitational force, acting towards the Earth, acts on the moving satellite. This is the **centripetal force** and it acts at 90° to the direction of motion of the satellite.
force (gravity)
velocity
The speed of the satellite remains constant, but the velocity constantly changes due to the change in direction. This means that the satellite is **accelerating** even though it is travelling at a constant speed.

Worked example

A motorcyclist is travelling around a circular curve in a flat road. What is providing the centripetal force? **(1 mark)**

☐ **A** gravity
☒ **B** the friction between the tyres and the road
☐ **C** the normal reaction of the road on the tyres
☐ **D** the thrust of the engine

The centripetal force must act at right angles to the direction of motion and **towards** the **centre of the circle**.

Centripetal forces

There are four main types of centripetal force that will result in circular motion. Here are some examples.

Force	Example
gravitational	orbiting satellite
frictional	car on a roundabout
tension	a hammer thrower
electrostatic	electron orbiting nucleus

Now try this

1 The figure shows an aircraft doing a circular loop.
(a) Draw arrows to show two forces acting on the aeroplane at point A. Label your arrows with the name of each force. **(2 marks)**
(b) Explain what happens to the velocity of the aircraft between point A and point B. **(2 marks)**

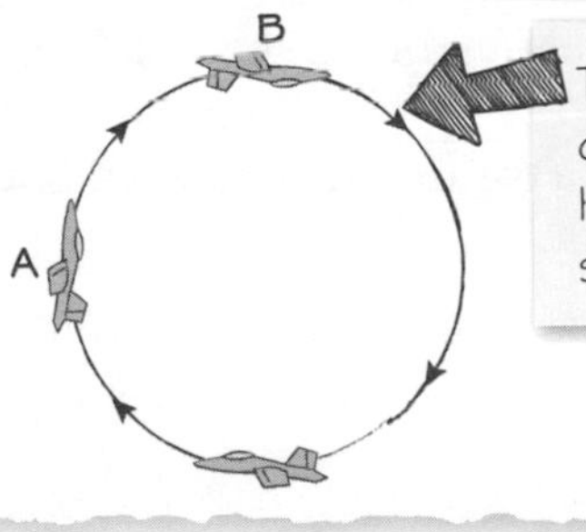

This is the motion of the aircraft in a horizontal circle, as seen from above.

Momentum and force

The momentum and the force of a moving body are closely related and can be understood by the use of Newton's second law of motion.

The **momentum** of a moving object depends on its mass and its velocity. Momentum is a **vector** quantity.

momentum (kg m/s) = mass (kg) × velocity (m/s)

$$p = m \times v$$

LEARN IT! IT'S NOT ON THE EQUATIONS LIST

Momentum is mass multiplied by velocity, so the units for momentum are a combination of the units for mass and velocity.

Momentum and force

Force is the rate of change of momentum.

Force (N) = change in momentum (kg m/s) ÷ time taken for the change (s)

$$F = \frac{mv - mu}{t}$$

Since $a = \frac{v - u}{t}$, then we also get the equation from Newton's second law, $F = ma$:

$$F = \frac{m(v - u)}{t}$$

Deceleration in cars and forces

When the mass or the deceleration of a vehicle are large, the forces exerted on the passengers are also large. This can be very dangerous.

Car safety features such as seat belts and crumple zones reduce the size of the forces on the driver and passengers by increasing the time over which the vehicle comes to rest.

Changing the size of the force

Changing the time changes the force needed to change the momentum.

By increasing the time over which the change in momentum takes place, the force needed is reduced.

$$\downarrow F = \frac{m(v - u)}{t \uparrow}$$

Worked example

A car of mass 1450 kg, travelling at 18 m/s, is brought to rest in 1.2 s. Calculate the force exerted on the car. **(4 marks)**

m = 1450 kg, u = 18 m/s, v = 0 m/s, t = 1.2 s

$$F = \frac{m(v - u)}{t}$$

F = 1450 kg (0 m/s – 18 m/s) ÷ 1.2 s

= –21 750 N

Worked example

A golfer wants to hit a golf ball with as much force as possible. Explain how the golfer does this. **(3 marks)**

The change in speed of the club and ball needs to be as high as possible and the contact time between the club and ball needs to be as small as possible. Both of these make the force as large as possible.

Now try this

1 Calculate the momentum of a 25 kg trolley moving at 6 m/s. **(3 marks)**

2 Calculate the force a car experiences when braking from 16 m/s to 7 m/s over a time period of 0.8 s when the car's mass is 1580 kg. **(3 marks)**

3 Explain how the size of the force exerted on passengers in a car crash can be reduced. **(3 marks)**

Newton's third law

Newton's third law relates to bodies in equilibrium and can be applied to collisions when considering the conservation of momentum.

Action and reaction

Newton's third law states that for every action there is an equal and opposite reaction.

The action force and the reaction force **act on different bodies**.

The man and wall are in equilibrium because the forces in the system are all balanced. There is no overall force in any direction.

Newton's third law examples

Watch out! Just because two forces are equal and opposite it does not always mean that they are an example of Newton's third law.

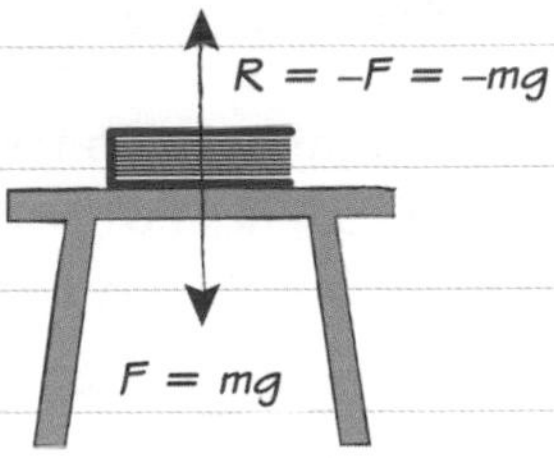

The reaction force of the table pushing up on the book, and the force of gravity acting downwards, are both acting on the same object – the book – so this is **not** an example of Newton's third law.

The gravitational pull of the Earth on the book and the gravitational pull of the book on the Earth **are** an example of Newton's third law, since they act on different objects, and are equal and opposite.

Newton's third law and collisions

When the truck and car collide, they exert equal and opposite forces on each other and are in contact for the same amount of time. Newton's third law is obeyed.

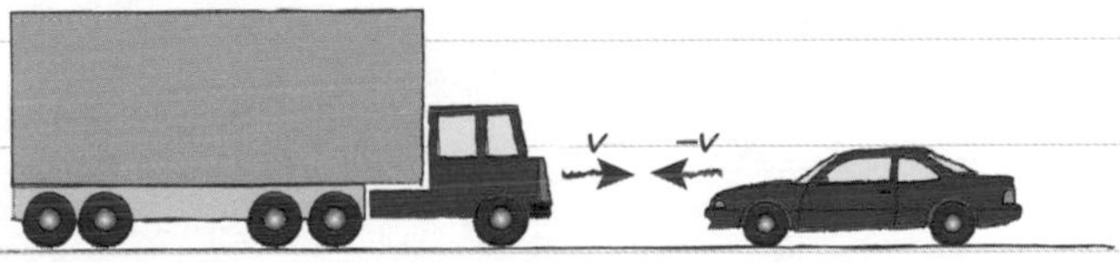

Momentum is conserved: the total momentum before the collision equals the total momentum after the collision.

Worked example

A model railway wagon (m) of mass 1 kg travelling at 12 m/s runs into a second stationary wagon (M) of mass 3 kg. After the collision the wagons stay linked together and move together. Calculate the velocity of the two wagons after the collision. **(4 marks)**

momentum = mass × velocity:

$1\text{ kg} \times 12\text{ m/s} = (1 + 3)\text{ kg} \times v$

$v = 12\text{ kg m/s} / 4\text{ kg}$

$v = 3\text{ m/s}$

The wagon with larger mass is stationary before the collision so has no momentum. The wagons move together after the collision with a combined mass of $(M + m)$.

Now try this

1 Explain how Newton's third law applies to collisions. **(2 marks)**

2 A truck of mass 12 kg moving at 8 m/s collides with a truck of mass 8 kg that is stationary. They collide and move off together. Calculate their shared velocity after the collision. **(4 marks)**

Human reaction time

Human reaction time is the time between a **stimulus** occurring and a **response**. It is related to how quickly the human brain can process information and react to it.

Human reaction time

It takes a typical person between 0.20 s and 0.25 s to react to a stimulus. Some people, such as international cricketers, 100 m sprinters and fighter pilots, train themselves to have improved reaction times.

For example, a top cricketer has a total time of 0.5 s to play a batting stroke. The first 0.2 s of this is the reaction time, the next 0.2 s is the batsman's preparation to play the shot and the final 0.1 s involves hitting the ball.

Reaction times and driving

Drivers have to react to changes in the traffic when driving. This may involve reacting to traffic lights changing colour, traffic slowing on a motorway or avoiding people or animals that may have walked in front of the vehicle.

The reaction time of humans may be affected by:

- tiredness
- distractions
- alcohol and drugs
- age.

Measuring reaction time

You can determine human reaction time by using the ruler drop test.

The reaction time is determined from the equation:

$$\text{reaction time} = \sqrt{\frac{2 \times \text{distance ruler falls}}{\text{gravitational field strength}}}$$

Repeats can be used to get a mean value for the reaction time.

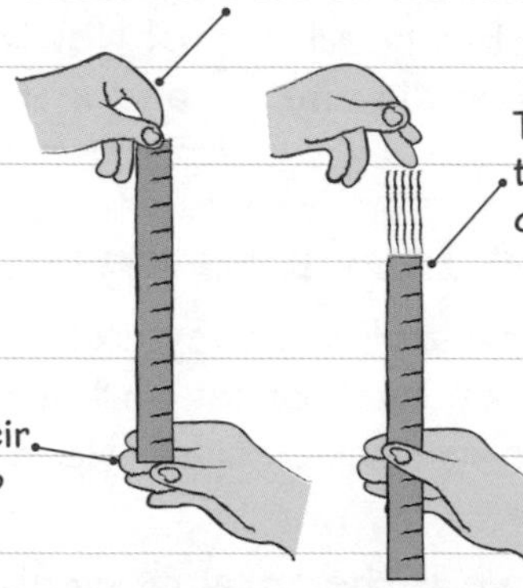

A metre ruler is held, by a partner, so that it is vertical and exactly level with the person's finger and thumb, with the lowest numbers on the ruler at the bottom.
The ruler is dropped and then grasped by the other person.
A person sits with their index finger and thumb opened to a gap of about 8 cm.

Worked example

A ruler drop test is conducted five times with the same person. The results show that the five distances fallen are: 0.16 m, 0.17 m, 0.18 m, 0.16 m and 0.15 m.

(a) Calculate the mean distance that the ruler drops. **(3 marks)**

mean = (0.16 + 0.17 + 0.18 + 0.16 + 0.15) ÷ 5 = 0.16 m

(b) Calculate the person's reaction time. **(3 marks)**

$$\text{reaction time} = \sqrt{\frac{2 \times 0.16}{10}} = 0.18\,\text{s}$$

Worked example

A driver moving at 30 km/h in her car has a reaction time of 0.25 s. Calculate the distance the car travels between seeing a hazard in the road and then applying the brakes. **(4 marks)**

distance (m) = speed (m/s) × time(s)

30 km/h = 30 000 m ÷ 3600 s = 8.3 m/s

distance travelled = 8.3 m/s × 0.25 s = 2.1 m

This distance for a driver is called the thinking distance. Read more about it on page 15.

Now try this

1 (a) Define human reaction time. **(1 mark)**

(b) Give three factors that can cause reaction times to increase. **(3 marks)**

2 Explain why the reaction time doubles when the distance that a ruler falls increases by a factor of 4. **(2 marks)**

Stopping distance

Stopping distance is the **total distance** over which a vehicle comes to rest.

It takes time for a moving car to come to a stop, and the car is still moving during this time. Understanding the factors that affect stopping distance is important for road safety.

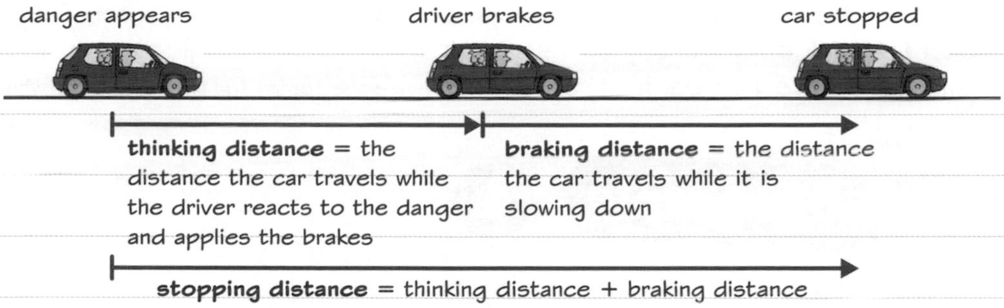
danger appears
driver brakes
car stopped
thinking distance = the distance the car travels while the driver reacts to the danger and applies the brakes
braking distance = the distance the car travels while it is slowing down
stopping distance = thinking distance + braking distance

Factors affecting stopping distance

Thinking distance and braking distance increase when the car's speed increases.

When you double the speed of a car, the thinking distance doubles, but the braking distance increases by a factor of four.

Thinking distance (driver's reaction time) increases when:

- the driver is **tired**
- the driver is **distracted**
- the driver has taken **alcohol** or **drugs**.

Braking distance increases when:

- the amount of **friction** between the tyre and the road decreases
- the road is **icy** or **wet**
- the brakes or tyres are **worn**
- the **mass** of the car is bigger.

Thinking distance and braking distance

Thinking distance is directly proportional to speed. If the thinking distance is 6 m when the car is travelling at 20 mph, then it will be 12 m when the car is travelling at 40 mph.

Braking distance is proportional to $(\text{speed})^2$.

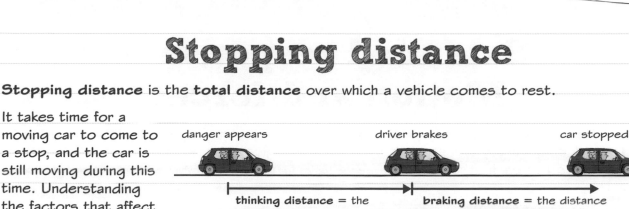

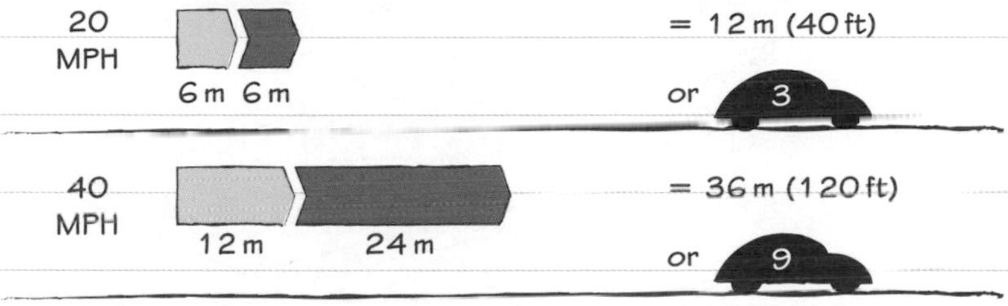
20 MPH
6 m 6 m
= 12 m (40 ft)
or 3
40 MPH
12 m 24 m
= 36 m (120 ft)
or 9

To bring a vehicle to rest, work must be done on it. When a force is applied to the brakes, the kinetic energy is transferred to thermal energy and the vehicle comes to rest over a certain distance.

The equation that governs this is:

$$F \times d = \frac{1}{2} \times m \times v^2$$

Worked example

A car is moving at a constant speed. Explain how the stopping distance changes when:

(a) the speed of the car increases **(2 marks)**

(b) the car has more passengers **(2 marks)**

(c) the driver has been drinking alcohol. **(2 marks)**

(a) The stopping distance increases because the thinking distance and braking distance increase.

(b) The mass of the car will be greater so the braking distance will increase, which means the stopping distance will increase.

(c) The stopping distance will increase because the thinking distance will increase.

Now try this

1 Explain how the thinking and braking distance of a car changes when the speed of the car increases from 20 mph to 80 mph **(2 marks)**

2 Calculate the braking distance of a car of mass 1250 kg travelling at 12 m/s if the average force applied to the brakes is 1800 N. **(3 marks)**

Use the equation $F \times d = \frac{1}{2} \times m \times v^2$

Extended response – Motion and forces

There will be one or more 6 mark questions on your exam paper. For these questions, you will need to think scientifically and structure your answer logically, showing how the points you make are related to each other. You can revise the topics for this question, which is about motion and forces, on pages 2–15.

Worked example

The diagram shows a trolley, ramp and light gates being used to investigate the relationship between the acceleration of a trolley on a slope and the angle that the slope makes with the horizontal.
Describe a simple investigation to determine how the acceleration of the trolley depends on the angle of the slope.

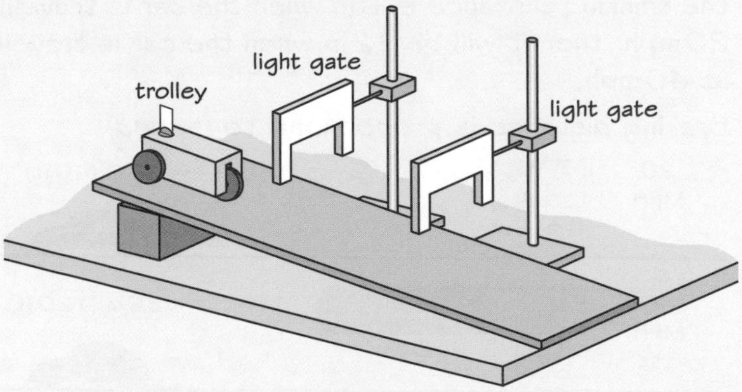

(6 marks)

I will start with the ramp at an angle of 10° and roll the trolley down the slope from the same starting position along the slope each time. The mass of the trolley needs to be kept constant, and a suitable value would be 500 g. I will then increase the ramp angle by 5° each time until I have at least 6 readings. I will repeat readings to check for accuracy and precision.

The speed of the trolley at each light gate is calculated using the equation speed = distance ÷ time, where the distance recorded is the length of the card that passes through the light gates. The acceleration of the trolley is calculated by finding the difference in the speed values at the two light gates and then dividing this by the time it takes the trolley to pass between the gates.

I would expect the acceleration of the trolley to increase as the angle increases, although the relationship may not be linear. This is because as the slope angle increases, you increase the component of the trolley's weight that is acting down the slope, which is the force responsible for the trolley's acceleration.

Command word: Describe

When you are asked to **describe** something, you need to write down facts, events or processes accurately.

Your answer should refer to the equations needed to calculate the velocity and acceleration of the trolley, as well as how you will show the relationship between acceleration and angle of slope.

You need to be careful when stating what you will vary and what you will keep the same. Since the dependent variable is acceleration and the independent variable is angle of slope, you will need to keep the mass of the trolley and its starting point the same each time.

Be clear when referring to how the speed of the trolley will be measured by referring to the card that passes through the light gates, and mention that the length of the card needs to be known if the light gates are recording time values. You need two speed and time values to calculate the acceleration.

You can also refer to the increase in gravitational potential energy as the slope angle increases. This means more is transferred to kinetic energy and so a greater change in velocity per second.

Now try this

The distance taken for vehicles to stop on roads depends on a number of factors. Discuss what these factors are and how they affect stopping distance.
(6 marks)

Energy stores and transfers

Energy can be stored in different ways, and can be transferred from one **energy store** to another.

Examples of energy stores

There are eight main energy stores.

Energy store	Example
chemical	fuel, food, battery
kinetic	moving objects
gravitational potential	raised mass
elastic	stretched spring
thermal	hot object
magnetic	two magnets
electrostatic	two charges
nuclear	radioactive decay

Closed systems

When there are **energy transfers** in a closed system, there is no net change to the total energy in the system.

A **closed system** is one where energy can flow in or out of the system, but there is no transfer of mass. An example of a closed system is a pan of water being heated that has a lid on it so that no steam can escape.

Energy transfers

Energy is transferred from one store to another in four different ways:

- **mechanically** – by a force moving through a distance
- **electrically** – by the use of electric current
- **thermally** – because of a difference in temperature
- **radiation** – by waves such as electromagnetic or sound.

Some of the energy will be usefully transferred, and some will be wasted or **dissipated**.

Examples of energy transfers

Energy transfers from one store to another can be shown on an energy flow diagram. This one shows a battery being used to lift a mass above the ground.

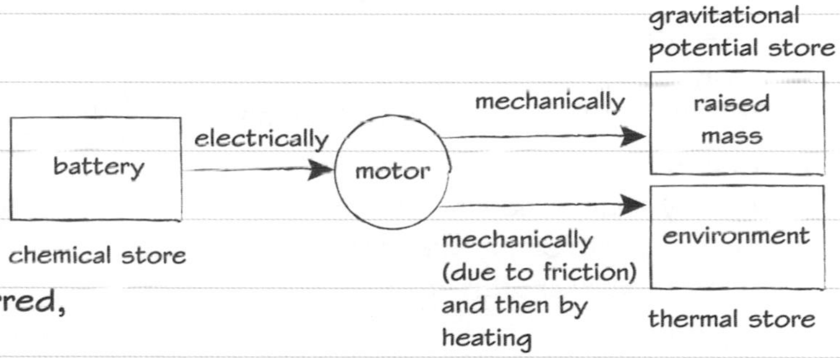

Mechanical processes become wasteful when they lead to a rise in the temperature of the surroundings through heating. In all system changes, energy is stored in less useful ways. For example, the light and sound energy from a TV will eventually be absorbed by the walls and by people, leading to a rise in their temperature.

Worked example

Draw a flow diagram to show the useful and wasted energy transfers for a camping gas stove. **(4 marks)**

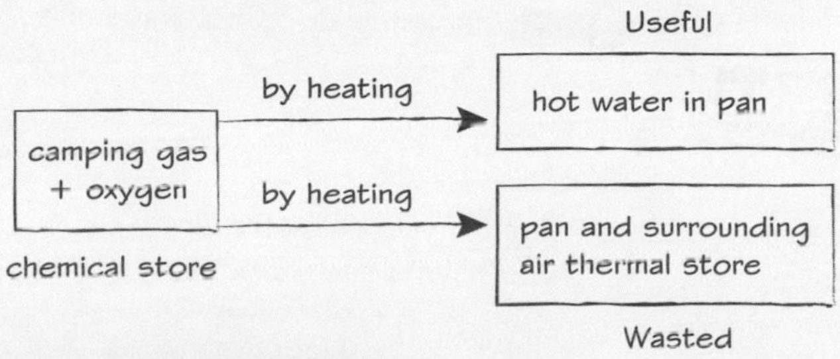

In all flow diagrams, there must be energy stores and energy transfers present to show how the energy is transferred from one store to others.

Based on the law of conservation of energy, the total energy present in the original stores must be equal to the energy present in the stores after the transfers have taken place, since you cannot create or destroy energy in any given system.

Now try this

1 Draw an energy store flow diagram for a car driving up a hill. **(4 marks)**

2 When a person lifts a 2 kg mass above the ground, the energy stores change. Draw a possible flow diagram for the transfer and explain how the energy stores change. **(5 marks)**

Efficient heat transfer

The rate at which a material transfers thermal energy depends on a number of factors. The **efficiency** of a device is a measure of how much **useful** energy it transfers.

Thermal energy transfer

The rate at which thermal energy is transferred through a wall of a house depends on:

1. the difference in temperature between the warmer interior and the colder exterior
2. the thickness of the walls
3. the material that the walls are made from.

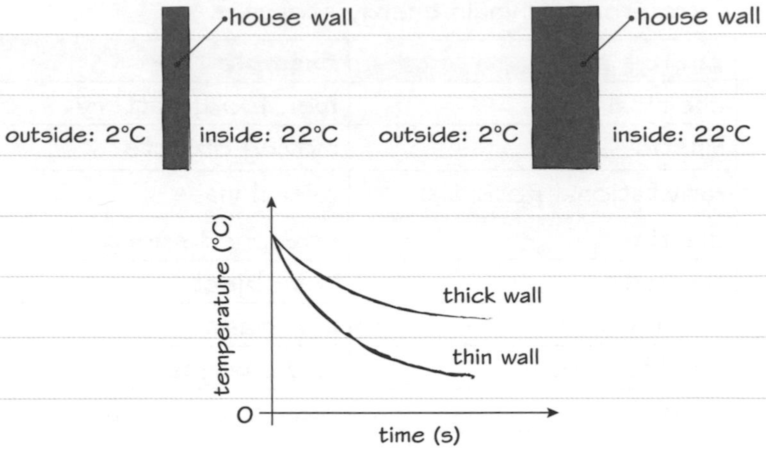

Efficiency

All machines **waste** some of the **energy** they **transfer**. Most machines waste energy as **heat** energy. The **efficiency** of a machine is a way of saying how good it is at transferring energy into **useful** forms.

A very efficient machine has an efficiency that is nearly 100%. The higher the efficiency, the better the machine is at transferring energy to useful forms.

Unwanted energy transfers can be reduced by thermal insulation and lubrication.

Thermal conductivity

A material with a high thermal conductivity is a better conductor of energy than one with a lower thermal conductivity. The rate at which the blue wall transfers energy is greater than that of the red wall. Different materials have different **relative thermal conductivities**.

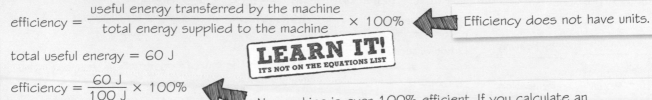

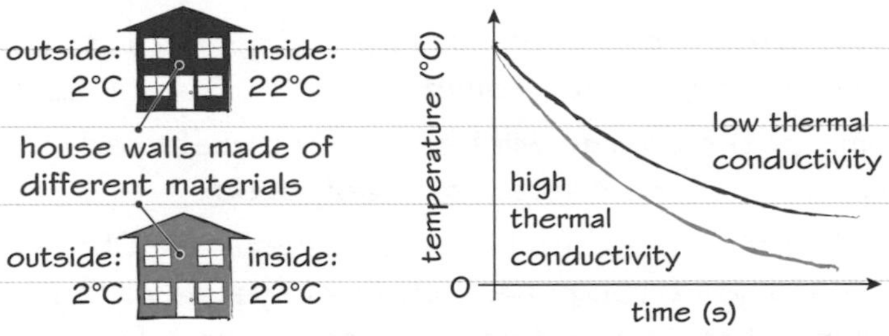

Worked example

A motor transfers 100 J of energy by electricity. 60 J are transferred as kinetic energy, 12 J as sound energy and 28 J as thermal energy. Calculate the efficiency of the motor. **(3 marks)**

$$\text{efficiency} = \frac{\text{useful energy transferred by the machine}}{\text{total energy supplied to the machine}} \times 100\%$$

Efficiency does not have units.

LEARN IT! IT'S NOT ON THE EQUATIONS LIST

total useful energy = 60 J

$$\text{efficiency} = \frac{60\text{ J}}{100\text{ J}} \times 100\%$$

$$= 60\%$$

No machine is ever 100% efficient. If you calculate an efficiency greater than 100% you have done something wrong!

Now try this

1 Calculate the efficiency of a lamp that transfers 14 J of energy into useful light energy for every 20 J of electrical energy input. **(2 marks)**

2 Explain why it is not possible for any device to be 100% efficient. **(3 marks)**

Use an electric appliance such as a kettle as an example.

Energy resources

The main energy resources include both **renewable** and **non-renewable** resources.

Renewable energy resources

Renewable energy resources will not run out. Most do not cause pollution or emit carbon dioxide.

1. **Bio-fuels** are animal or plant matter used to produce thermal energy, electrical energy or used to power cars.
2. **Hydroelectricity** generates electricity from water behind a dam flowing down a pipe and turning a turbine to generate electricity.
3. **Wind turbines** use kinetic energy from the wind to generate electricity.
4. **Tidal power** uses the rise and fall of the tide or tidal currents to generate electricity.
5. **Solar cells** convert solar energy, or energy from the Sun, directly to electricity. Solar energy can also be used directly to cook food or to heat water.

Non-renewable energy resources

Non-renewable resources are resources that will run out eventually.

1. Nuclear fuels such as uranium are used:
   - to generate electricity
   - as energy sources in spacecraft.
2. Fossil fuels are coal, oil and natural gas. They are used:
   - to generate electricity
   - to power transport
   - to heat homes and for cooking.
3. They are available all the time, unlike some renewable resources.

Worked example

Describe how different renewable and non-renewable resources are used for transport. **(4 marks)**

Renewable and non-renewable energy resources can both be used in power stations, leading to the production of electricity which is used to power trains. Electrical energy can also be supplied from other renewable sources such as solar, hydroelectric and wind.

Biofuel, in the form of methanol, is an example of how renewable fuel can be used to power cars, and solar power is also being developed to do more of this. Petrol and diesel, from oil, are examples of non-renewable fuels that are used to power cars.

Now try this

1 Draw diagrams to show the energy transfers that take place
(a) in hydroelectricity **(3 marks)**
(b) when biofuels are used for transport. **(3 marks)**

2 Suggest the advantages and disadvantages of using biofuels to power cars compared with using solar power to power cars. **(4 marks)**

Patterns of energy use

The patterns and trends in how humans have used energy resources have changed over time.

Patterns and trends in energy use

The use of energy resources has changed over the years. This is because of several factors.

1. **the world's population** In the past 200 years, the population of the world has risen from 1 billion to 7 billion people.
2. **the development of technology** Vehicles such as cars, trains and planes and other devices have increased in number and these all require energy.
3. **electrical energy** Power stations require fuels in order to generate electricity.

Growth in world population

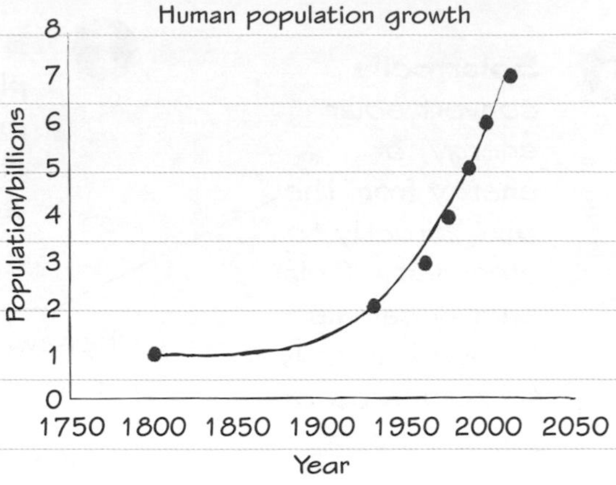

Trends in the world's energy use

As the graph shows, most of the world's energy use has been fossil fuels. Before about 1900, biomass in the form of wood was the major source of fuel and its use has remained constant over time.

In more recent years, there has been an increase in the use of nuclear fuel and hydroelectric power.

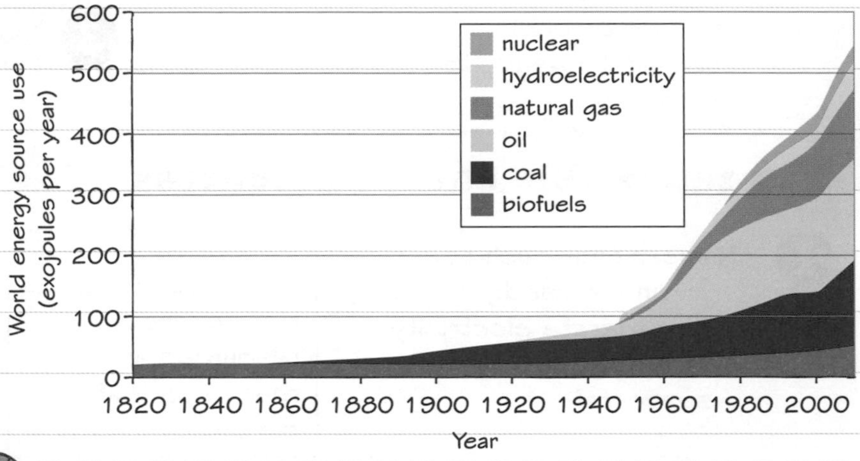

Worked example

Compare the shape of the world's population graph with that of the energy use graph. **(4 marks)**

They are similar in that as the population increases, the total energy used will increase, assuming that each person uses the same amount of energy. They differ in that the total energy use has increased faster than the population. As societies develop, they use more technology that requires energy to run it. This means that the average amount of energy used per person also increases.

The future?

It is not possible to continue using the Earth's non-renewable energy reserves to the extent that they are being used now. They are a finite reserve, so will run out and not be replaced.

Greater use of fossil fuels will lead to more carbon dioxide in the atmosphere. There will be greater global warming, leading to severe weather, flooding and threats to food supplies.

Now try this

1 Describe the main issues with using the Earth's energy supplies. **(2 marks)**

2 Wood is a biofuel. Suggest why wood has been used constantly for many years, whereas other fuels have only been used much more recently. **(3 marks)**

Potential and kinetic energy

You can calculate the **kinetic** and **gravitational potential energy** of objects using the equations on this page.

Gravitational potential energy

Gravitational potential is the energy possessed by a body due to its height above the Earth. The value of the gravitational potential energy stored depends on:

- the mass of the body
- the gravitational field strength
- the height the body is raised.

The change in gravitational potential energy is given by the equation:

change in gravitational potential energy (J) = mass (kg) × gravitational field strength (N/kg) × change in vertical height (m)

$$\Delta GPE = m \times g \times \Delta h$$

LEARN IT! IT'S NOT ON THE EQUATIONS LIST

The gravitational field strength on Earth is 10 N/kg. It can also be given in m/s^2.

Kinetic energy

Kinetic energy is stored in moving objects and is calculated using the equation:

kinetic energy (J) = $\frac{1}{2}$ × mass (kg) × $(\text{speed})^2$ $(m/s)^2$

$$KE = \frac{1}{2} \times m \times v^2$$

LEARN IT! IT'S NOT ON THE EQUATIONS LIST

Maths skills Kinetic energy is directly proportional to the mass of the moving object: doubling the mass doubles the kinetic energy of the moving object.

It is directly proportional to the square of the speed, so doubling the speed means the KE increases by a factor of four.

Worked example

(a) A mass of 800 g is moving at 14 m/s. Calculate its kinetic energy. **(3 marks)**

$KE = \frac{1}{2} \times m \times v^2$

$KE = \frac{1}{2} \times 0.8 \times 14^2 = 78.4$ J

(b) A body of mass 4.8 kg has kinetic energy of 200 J. Calculate the speed it is moving at. **(4 marks)**

$200\text{ J} = \frac{1}{2} \times 4.8 \times v^2$

Rearranging gives $v = \sqrt{400 \div 4.8} = 9.1$ m/s

Remember to square the value of speed when calculating the kinetic energy. The term v^2 means $v \times v$.

Worked example

(a) A body of mass 73 kg is lifted through a vertical height of 26 m. Calculate how much gravitational potential energy it has gained. **(3 marks)**

$\Delta GPE = m \times g \times \Delta h$

$= 73\text{ kg} \times 10\text{ N/kg} \times 26\text{ m} = 18980\text{ J}$

(b) The body is now dropped from 26 m above the ground. At what speed will it hit the floor? **(4 marks)**

using the law of conservation of energy:

loss in GPE = gain in KE

$18980\text{ J} = \frac{1}{2} \times 73\text{ kg} \times v^2$

rearranging gives $v^2 = 520\,(m/s)^2$

hence $v = \sqrt{520} = 23$ m/s

Now try this

1 A mass of 5 kg is raised through a vertical height of 18 m. Calculate the change in gravitational potential energy. **(3 marks)**

2 A motorbike of mass 80 kg is moving at 30 km/h. Calculate its kinetic energy. **(4 marks)**

3 A ball of mass 3 kg falls from 34 m above the ground. Calculate its speed when it lands. **(4 marks)**

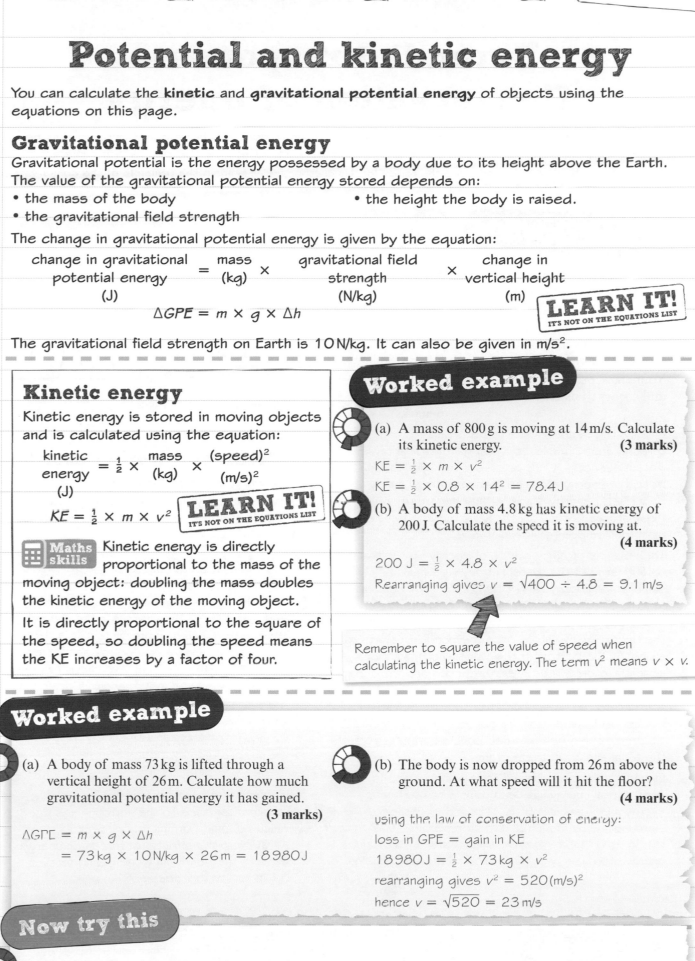

Extended response – Conservation of energy

There will be one or more 6 mark questions on your exam paper. For these questions, you will need to think scientifically and structure your answer logically, showing how the points you make are related to each other. You can revise the topics for this question, which is about **the principle of the conservation of energy**, on pages 17–20.

Worked example

Figure 1 shows the arrangement of apparatus for a simple pendulum. Figure 2 shows how the kinetic energy of the pendulum changes with time.

Describe the energy transfers taking place during the motion of the simple pendulum.

Your answer should refer to the energy stores and energy transfers that are involved.

(6 marks)

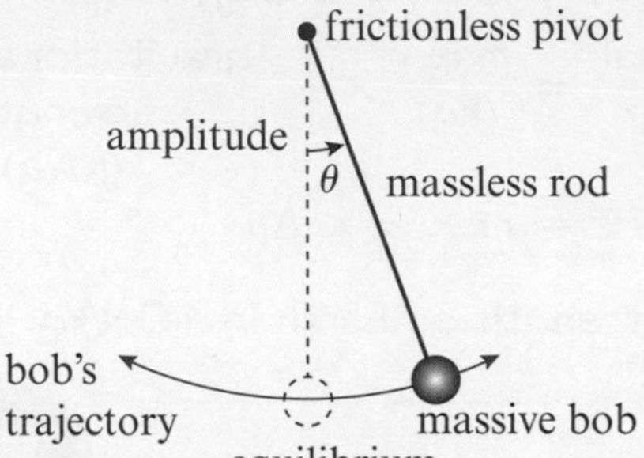

Figure 1

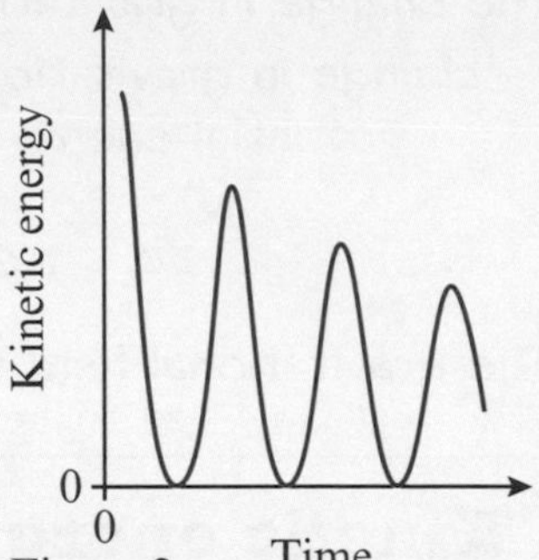

Figure 2

Initially, the pendulum has to be raised through a small angle, so it will gain gravitational potential energy. Once released, this gravitational potential energy store will be transferred mainly to a kinetic energy store, with some energy being dissipated as sound and heat by thermal transfer from friction due to air resistance.

The principle of conservation of energy states that the total energy of the system at any time must be the same, since energy cannot be created nor destroyed. Initially, this was all in the gravitational potential energy store, but this is then mechanically transferred to the kinetic energy store once the pendulum starts to swing. At the bottom of the swing, the gravitational potential store will be greatly reduced and the energy will be mostly in the kinetic store. Some energy is also transferred to the surroundings as heat and sound by thermal transfer from friction due to air resistance, which must also be taken into consideration as part of the total energy of the system.

Eventually, the pendulum will stop swinging, so the gravitational energy has been transferred to kinetic energy and it has eventually all ended up in the surroundings by friction due to air resistance, leading to the surroundings becoming slightly warmer. The useful energy store has been dissipated to the surroundings.

Command word: Describe

When you are asked to **describe** something, you need to write down facts, events or processes accurately.

It is a good idea to state what the principle of conservation of energy is, so that the examiner knows that you understand it. You can then apply this idea to the changes in the energy stores and transfers in the system. It is also worth noting that the way we talk about energy at GCSE now has changed – you need to refer to energy stores as well as energy transfers. In this example, the main stores are gravitational and kinetic, and the motion of the pendulum is via mechanical transfer, with energy transfer to the surroundings by thermal transfer.

In energy transfers, the energy stores are usually dissipated to the surroundings, which become warmer. In a perfect situation, with no frictional forces or energy wasted, the pendulum would swing forever, but in reality this does not happen.

Refer to *energy stores* and *energy transfers* in your answer. There are eight energy stores and four energy transfers that you need to know. See page 17 to remind yourself of these.

Now try this

The people on a small island want to use either wind power, on hills and out at sea, or its reserves of coal to generate electrical energy. Compare the advantages and disadvantages of each resource. **(6 marks)**

Waves

Waves transfer energy and information without transferring matter. Evidence of this for a water wave can be seen when a ball dropped into a pond bobs up and down, but the wave energy travels outwards as ripples across the surface of the pond.

Waves can be described by their

- **frequency** – the number of waves passing a point each second, measured in **hertz (Hz)**
- **speed** – measured in **metres per second (m/s)**
- **wavelength** and **amplitude**
- period – the time taken for one wavelength to pass a point
- **period** = 1/frequency

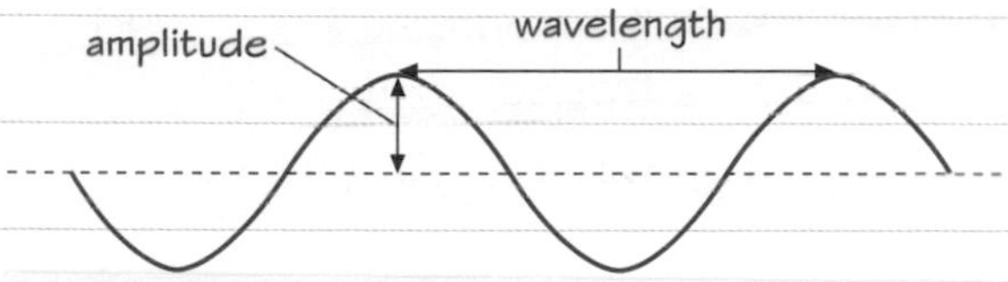

Watch out! Remember that the amplitude is *half* of the distance from the top to the bottom of the wave. Students often get this wrong.

Longitudinal waves

Sound waves and **seismic P waves** are **longitudinal** waves. The particles in the material the sound is travelling through move back and forth along the same direction that the sound is travelling.

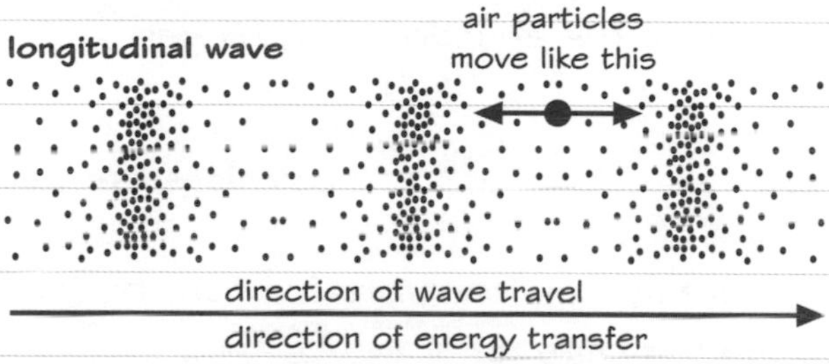

Particles in a **longitudinal** wave move **along** the same direction as the wave.

Transverse waves

Waves on a **water surface, electromagnetic waves** and **seismic S** waves are all **transverse** waves. The particles of water move in a direction at right angles to the direction the wave is travelling.

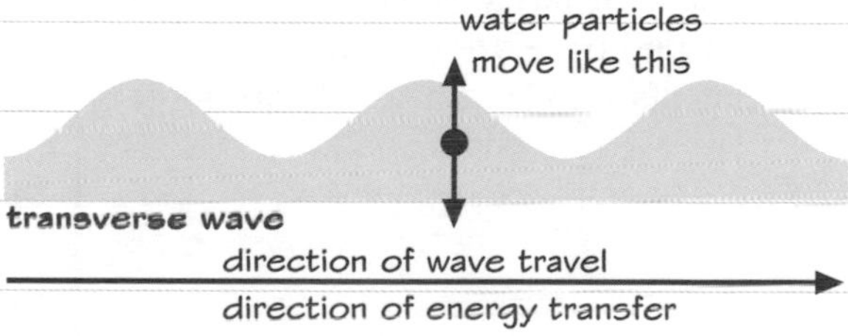

Particles in a **transverse** wave move **across** the direction the wave is travelling.

Worked example

Give two ways in which longitudinal and transverse waves are

(a) similar **(2 marks)**

They both transfer energy without transferring matter and have an amplitude, speed, wavelength and frequency.

(b) different. **(2 marks)**

Particles in longitudinal waves vibrate along the direction of movement, whereas particles in transverse waves move at 90 degrees to the direction of travel. They can also have different speeds, frequencies and wavelengths.

Now try this

1 (a) Sketch a transverse wave and mark the amplitude and wavelength on it. **(3 marks)**

(b) Draw an arrow to show which way the wave moves. **(1 mark)**

(c) Draw a small particle on the wave, with arrows to show which way it moves. **(1 mark)**

2 The graph shows a wave. Each vertical square represents 1 mm. Work out the amplitude of the wave. **(1 mark)**

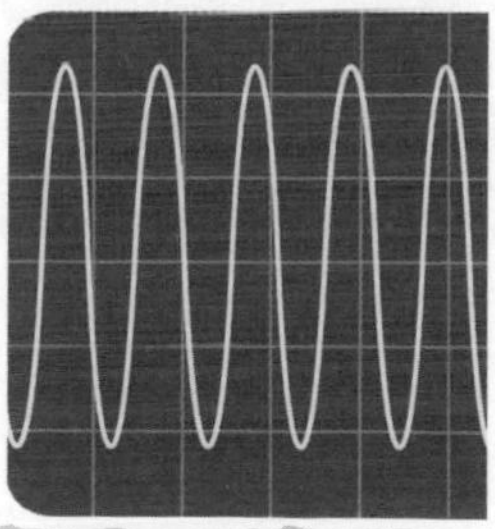

Wave equations

Both of these equations can be use to find the **speed** or **velocity** of a wave.

Speed, frequency and wavelength

LEARN IT! IT'S NOT ON THE EQUATIONS LIST

$v = f \times \lambda$

v = wave speed (metres per second, m/s)

f = frequency (hertz, Hz)

λ = wavelength (metres, m)

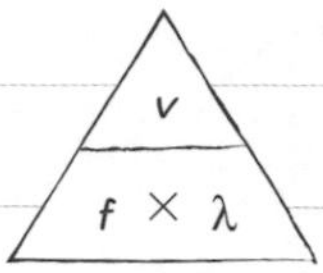

Speed, distance and time

LEARN IT! IT'S NOT ON THE EQUATIONS LIST

$v = \frac{x}{t}$

v = wave speed (metres/second, m/s)

x = distance (metres, m)

t = time (seconds, s)

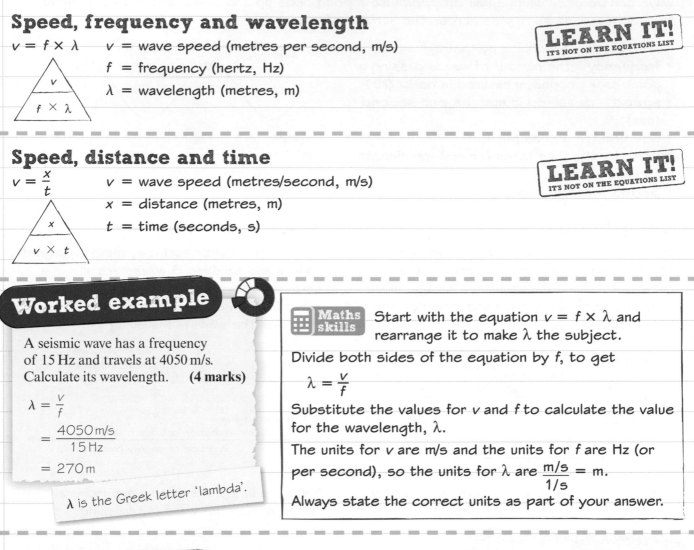

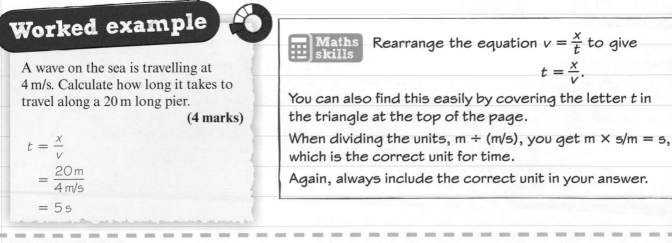

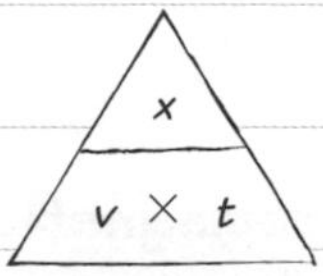

Worked example

A seismic wave has a frequency of 15 Hz and travels at 4050 m/s. Calculate its wavelength. **(4 marks)**

$\lambda = \frac{v}{f}$

$= \frac{4050\,\text{m/s}}{15\,\text{Hz}}$

$= 270\,\text{m}$

λ is the Greek letter 'lambda'.

Maths skills Start with the equation $v = f \times \lambda$ and rearrange it to make λ the subject.

Divide both sides of the equation by f, to get

$\lambda = \frac{v}{f}$

Substitute the values for v and f to calculate the value for the wavelength, λ.

The units for v are m/s and the units for f are Hz (or per second), so the units for λ are $\frac{\text{m/s}}{1/\text{s}}$ = m.

Always state the correct units as part of your answer.

Worked example

A wave on the sea is travelling at 4 m/s. Calculate how long it takes to travel along a 20 m long pier. **(4 marks)**

$t = \frac{x}{v}$

$= \frac{20\,\text{m}}{4\,\text{m/s}}$

$= 5\,\text{s}$

Maths skills Rearrange the equation $v = \frac{x}{t}$ to give $t = \frac{x}{v}$.

You can also find this easily by covering the letter t in the triangle at the top of the page.

When dividing the units, m ÷ (m/s), you get m × s/m = s, which is the correct unit for time.

Again, always include the correct unit in your answer.

Now try this

1 A sound wave with a frequency of 100 Hz has a speed of 330 m/s. Calculate its wavelength. **(3 marks)**

2 A wave in the sea travels at 25 m/s. Calculate the distance it travels in one minute. **(4 marks)**

Measuring wave velocity

You need to be able to **calculate** the speed of sound in air or the speed of ripples on the surface of water.

Calculating the speed of sound in air

Method 1: using an echo.

1. Measure the distance from the source of the sound to the reflecting surface (the wall).
2. Measure the time interval, with a stopwatch, between the original sound being produced and the echo being heard.
3. Use speed (m/s) = distance (m) ÷ time (s) to calculate the speed of sound in air.

Repeating the experiment a number of times over a range of distances will allow you to obtain accurate and precise results.

Method 2: using two microphones and an oscilloscope.

1. Set up the microphones one in front of the other at different distances in a straight line from a loudspeaker.
2. Set the frequency of the sound from the loudspeaker to a known, audible value.
3. Display the two waveforms on the oscilloscope. Measure the distance between the microphones.
4. Move the microphones apart so that the waveforms move apart by 1 wavelength.
5. Calculate the speed of sound using the equation: wavespeed (m/s) = frequency (Hz) × wavelength (m)

Calculating the speed of ripples on water surfaces

You can work out the speed of ripples on the surface of water using a ripple tank and a strobe.

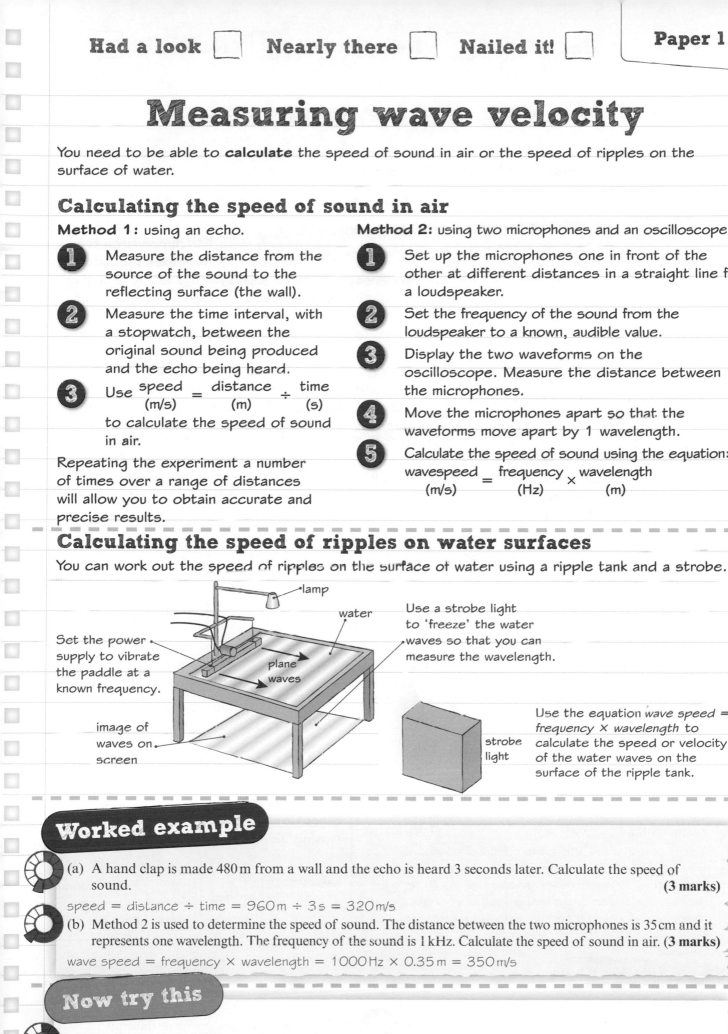

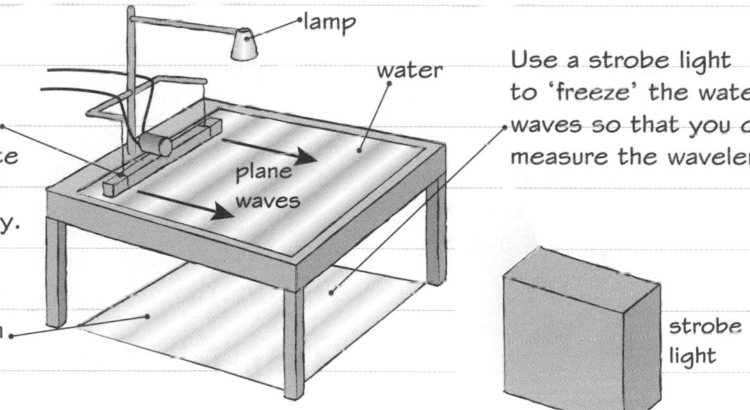

lamp
water
Set the power supply to vibrate the paddle at a known frequency.
plane waves
Use a strobe light to 'freeze' the water waves so that you can measure the wavelength.
image of waves on screen
strobe light

Use the equation *wave speed* = *frequency* × *wavelength* to calculate the speed or velocity of the water waves on the surface of the ripple tank.

Worked example

(a) A hand clap is made 480 m from a wall and the echo is heard 3 seconds later. Calculate the speed of sound. **(3 marks)**

speed = distance ÷ time = 960 m ÷ 3 s = 320 m/s

(b) Method 2 is used to determine the speed of sound. The distance between the two microphones is 35 cm and it represents one wavelength. The frequency of the sound is 1 kHz. Calculate the speed of sound in air. **(3 marks)**

wave speed = frequency × wavelength = 1000 Hz × 0.35 m = 350 m/s

Now try this

1 The frequency of a ripple tank paddle is 4 Hz and the wavelength of the ripples is 8 cm. Calculate the speed of the water waves in m/s. **(3 marks)**

2 Explain how you could obtain a value for the speed of sound in air with a small percentage error. **(4 marks)**

Waves and boundaries

Waves can show different effects when they move from one material to another. These changes can occur at the **boundary** or **interface** between the two materials.

Waves and boundaries

Whenever a sound wave, light wave or water wave reaches the boundary between two materials, the wave can be:

- **reflected**
- **transmitted** or
- **refracted**
- **absorbed**.

Different substances reflect, refract, transmit or absorb waves in ways that vary with the wavelength.

Different wavelengths of radiation are absorbed by molecules in the atmosphere by different amounts.

Sound waves at a boundary

The amount of reflection that happens at the boundary between two materials depends on the densities of the materials. The greater the difference in density, the more sound energy will be reflected.

- Sound is **transmitted** through a material when the **densities** are similar.
- Sound can be **absorbed** by materials. The amount of absorption depends on the material and the **wavelength** of the sound.
- Sound is **reflected** when there is a **big difference in the densities** of the materials at an interface, for example between air and concrete.

Refraction

Sound waves, water waves and light waves can all be refracted. Refraction can result in a change of both speed and direction. The direction does not change if the wavefronts travel perpendicular to the normal.

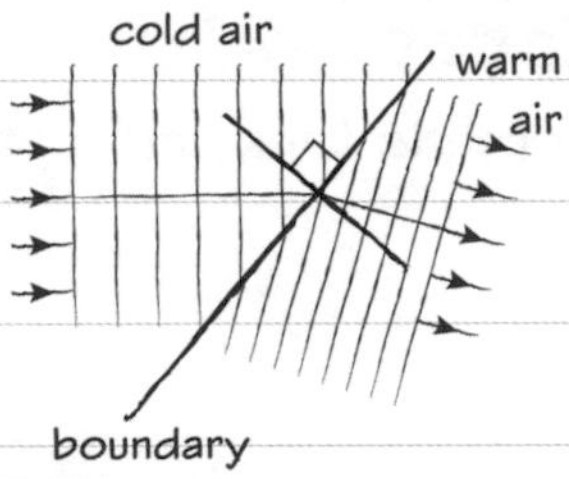

Sound waves travel slower in cooler, denser air than in warmer, less dense air.

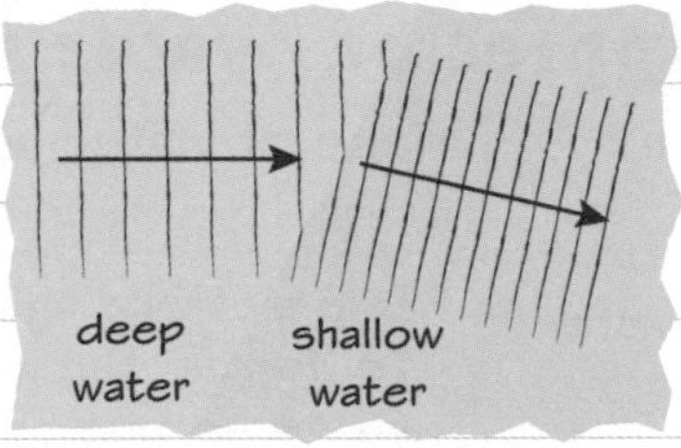

Water waves travel faster in deep water than in shallow water. They can also change direction.

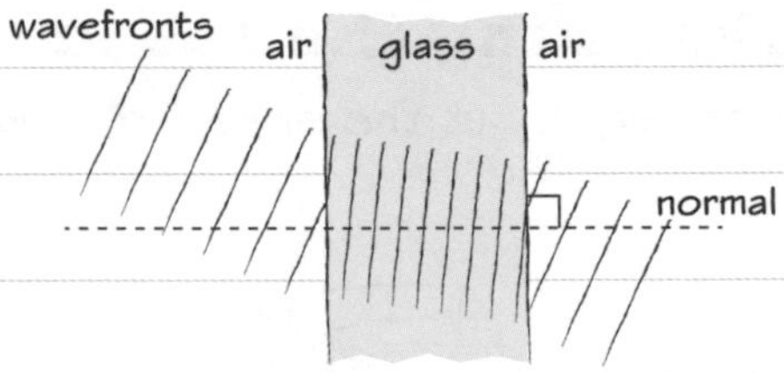

Light waves can slow down and change direction when they pass from air to glass.

Worked example

A person blows a dog whistle. The dog is 200 m away. Explain why the dog will hear the whistle sooner on a warm day than on a cool day. **(3 marks)**

Air temperature will affect the speed of sound. Air molecules at a higher temperature have more kinetic energy than air molecules at a lower temperature, so they vibrate faster. Since the air molecules vibrate faster, sound waves travel through the warmer air more quickly than through the cooler air.

Refraction special case

When light, sound or water waves move from one material into another their **direction does not change** if they are moving along the normal.

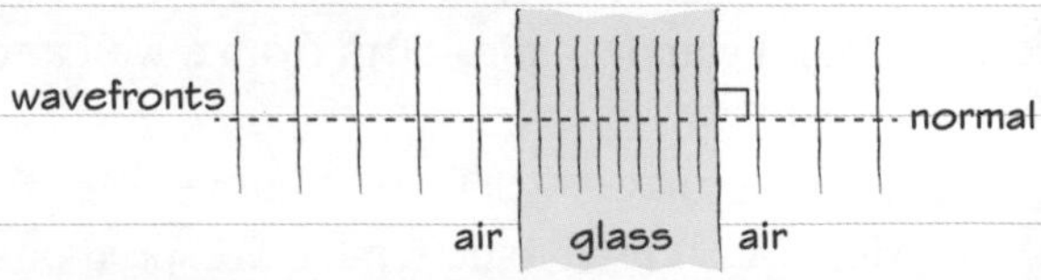

Now try this

1 State four things that can happen to a wave at a boundary between two materials. **(2 marks)**

2 Explain the changes that occur to a wave during refraction. **(3 marks)**

3 Suggest how the absorption of different wavelengths of infrared radiation by molecules in the atmosphere is
(a) useful **(2 marks)**
(b) a problem. **(2 marks)**

Sound waves and the ear

Waves produce **vibrations** in solids and this is the process used by the ear to detect sound waves.

Vibrations and waves

When objects vibrate, sound waves are produced. These sound waves are a series of oscillations that transfer energy from the source to the ear. One example of this is the vibration of a guitar string to produce a musical note. The length, thickness and material of the string determines the frequency of the note. Once plucked, the string will vibrate at its natural frequency.

The human ear can only detect frequencies in the range 20 Hz to 20 kHz. It does not hear frequencies below 20 Hz or above 20 kHz.

Energy transfer

Air molecules are forced to vibrate by the source of the vibration. Energy travels via a longitudinal wave until it reaches the ear.

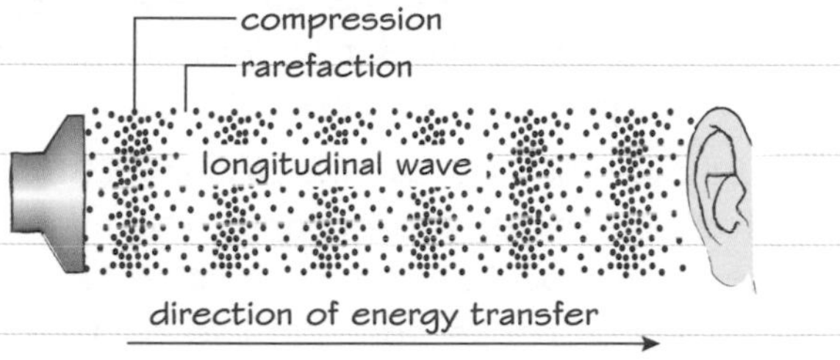

How sound waves are produced

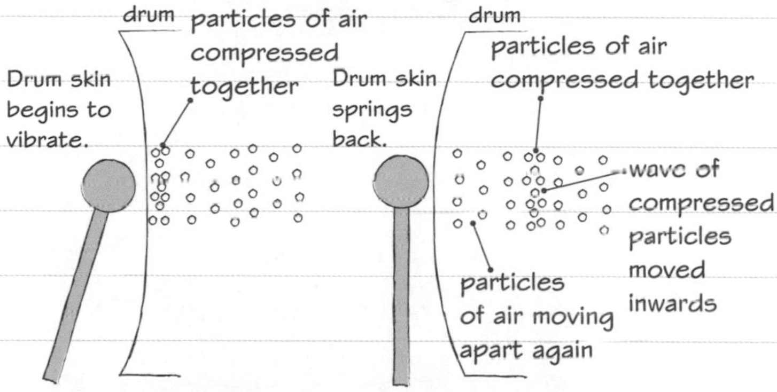

After the drum is hit, the vibration of the drum will cause a longitudinal wave to move outwards from it.

How the ear detects sound

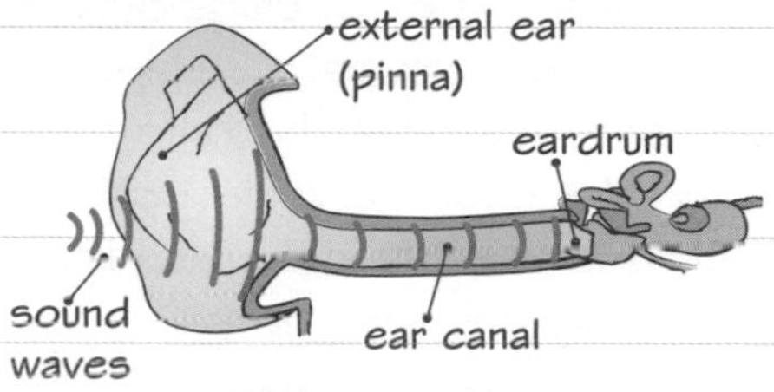

Sound waves are channelled down the ear canal and cause the eardrum to vibrate. These vibrations pass through the ear as further vibrations are then converted to an electrical signal and carried to the brain.

Worked example

Explain why the human ear can only hear sounds in the range from 20 Hz to 20 000 Hz. **(2 marks)**

Sound waves travelling from air into a solid are converted to vibrations and travel through the solid as a series of vibrations. This conversion of sound waves to vibrations in solids only works over a limited frequency range, which is from 20 Hz to 20 kHz.

As you get older, the range of frequencies you can hear decreases. Older people may only be able to hear up to 12 kHz or even lower. Hearing loss can be caused by exposure to loud sounds or to sound over a long period of time.

The frequency of vibration of a solid is the same as the frequency of the sound wave that is causing it to vibrate. Solids can be made to vibrate by these waves, but not for every frequency of the wave. So, the eardrum will not vibrate if the wave frequency is less than 20 Hz or more than 20 kHz.

Now try this

1 Explain what the ear does in terms of the energy transfers that take place inside the ear. **(3 marks)**

2 Explain how a vibration becomes a sound that is heard. **(3 marks)**

3 Explain why sound waves are often described by scientists in terms of their frequencies rather than their wave speed. **(4 marks)**

Ultrasound and infrasound

Ultrasound is sound waves with a frequency above 20000 Hz (20 kHz). **Infrasound** is sound waves with a frequency lower than 20 Hz.

Exploring the Earth's core

Infrasound is believed to travel through the earth as shockwaves from tsunami, volcanoes or earthquakes. Infrasound is also produced when meteors enter the Earth's atmosphere. Scientists can detect this infrasound and track the path the meteor will take.

Infrasound waves can be detected from explosions under the ground. The infrasound waves can also help to determine the structure of rocks beneath the earth's crust.

Pre-natal scans

Ultrasound waves are used to make images of the inside of the body. Ultrasounds are not harmful, so it is safe to use them to scan **foetuses** (unborn babies). The ultrasound waves are sent into the woman's body, and some of the sound is reflected each time it meets a layer of tissue with a different density to the one it has just pased through. The scanner detects the echoes and a computer uses the information to make a picture.

Sonar

Sonar uses pulses of ultrasound to find the depth of water beneath a ship. The sonar equipment measures the time between sending the sound and detecting its echo. This time is used to calculate the depth of the water, using this equation:

distance = speed × time

(see page 3)

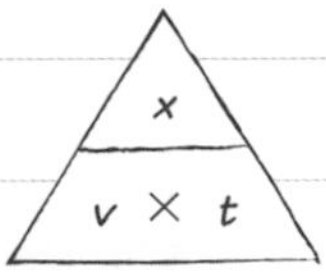
x
v × t

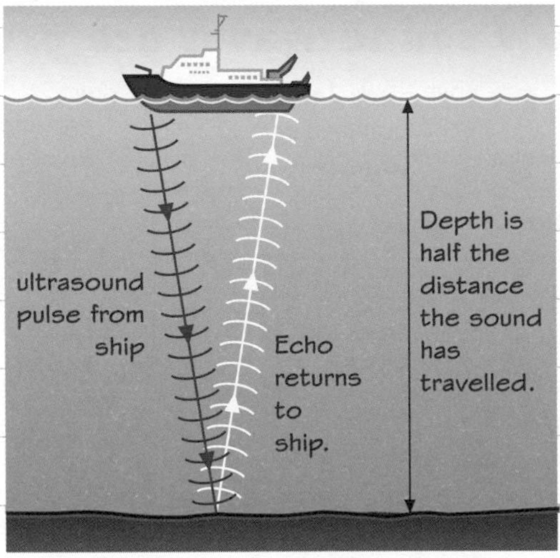
ultrasound pulse from ship
Echo returns to ship.
Depth is half the distance the sound has travelled.

Worked example

A ship detects an echo 5 seconds after it has sent out a sonar pulse. Sound travels at 1500 m/s in sea water. How deep is thc water? **(4 marks)**

distance = speed × time
= 1500 m/s × 5 sec
= 7500 m

depth of water = $\frac{7500\text{ m}}{2}$
= 3750 m

Be careful with the units. If the speed is in metres per second, the time must be in seconds.

Remember that the depth of the water is *half* the distance the ultrasound has travelled.

Now try this

1 Describe the similarities and differences between sound, infrasound and ultrasound. **(3 marks)**

2 Describe how ultrasound waves can be used to produce an image of a foetus. **(4 marks)**

3 A sonar signal is sent from a ship. The echo is received 0.2 s later, back at the ship. The sound wave has travelled a total distance of 284 m. Calculate the speed of sound in water. **(3 marks)**

Sound wave calculations

You can calculate the changes in **wavelength** and **velocity** of a sound wave when it moves from one medium to another.

Changes in sound waves

When a sound wave moves from one material to another:

- the wave speed or **velocity** may change
- the **wavelength** may change
- the **frequency** will not change.

$$\text{wave speed (m/s)} = \text{frequency (Hz)} \times \text{wavelength (m)}$$

$$v = f \times \lambda$$

Wavelength, λ, is **directly proportional** to wave speed, v.

If the speed or velocity of a sound wave increases, then its wavelength will also increase. However, its frequency does not change as the number of waves being produced per second is not affected.

Refraction of sound waves

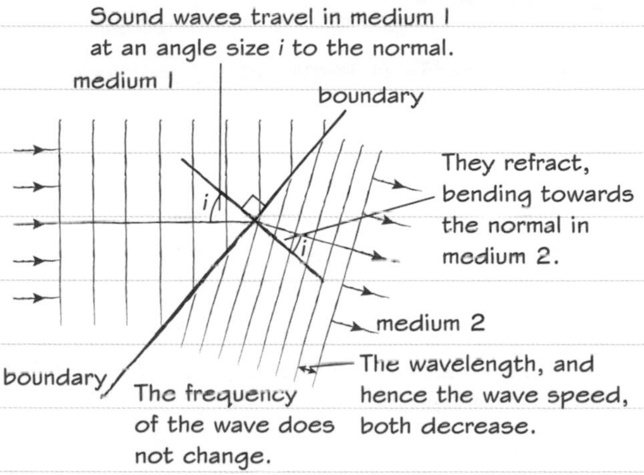
Sound waves travel in medium 1 at an angle size i to the normal.
medium 1
boundary
They refract, bending towards the normal in medium 2.
i
i
medium 2
The wavelength, and hence the wave speed, both decrease.
boundary
The frequency of the wave does not change.

Typical sound wave speeds

Material	Density (g/cm³)	Speed of sound (m/s)
steel	7.86	5940
water	1.00	1496
fat	0.95	1450
muscle	1.07	1580
air	0.001 39	331

Speed of sound

The speed of sound in a material usually depends on the **density** of the material. The denser the material, the greater the speed, since the wave can be passed more easily from particle to particle in a dense material.

Solids are usually denser than liquids, with gases being the least dense materials.

Worked example

A wave of frequency 400 Hz has a wavelength of 0.8 m.

(a) Calculate the speed of the wave in m/s. **(3 marks)**

speed = frequency × wavelength
= 400 Hz × 0.8 m = 320 m/s

(b) Calculate its speed when it enters a material where its wavelength increases to 2.5 m. **(3 marks)**

Wavelength has increased from 0.8 m to 2.5 m, and the frequency has stayed the same.

new wave speed = 400 Hz × 2.5 m = 1000 m/s

Worked example

Explain why sound waves travel at different speeds in different states of matter. **(3 marks)**

The speed of sound is related to the density of the material. The greater the density of the material, the faster the sound wave will travel through it, since the particles will be more closely packed, allowing energy to be transferred between vibrating particles more easily.

Now try this

1 Explain why sound waves do not travel through a vacuum. **(2 marks)**

2 (a) The ratio of the wavelengths of a sound wave that has passed across the boundary between two materials is 2 : 5. The speed of sound in the material of lower density is 342 m/s. Calculate the wave's speed in the material of higher density. **(2 marks)**

(b) State the assumptions you have made. **(2 marks)**

Waves in fluids

Practical skills You can **determine the wave speed, frequency and wavelength** of waves by using appropriate apparatus.

Core practical

Investigating waves

Aim

To investigate the suitability of apparatus to measure the speed, frequency and wavelength of waves in a fluid.

Apparatus

ripple tank, motor, plane wave generator, stroboscope, ruler, A3 paper and pencil

Method

1 Set up the apparatus as shown below.

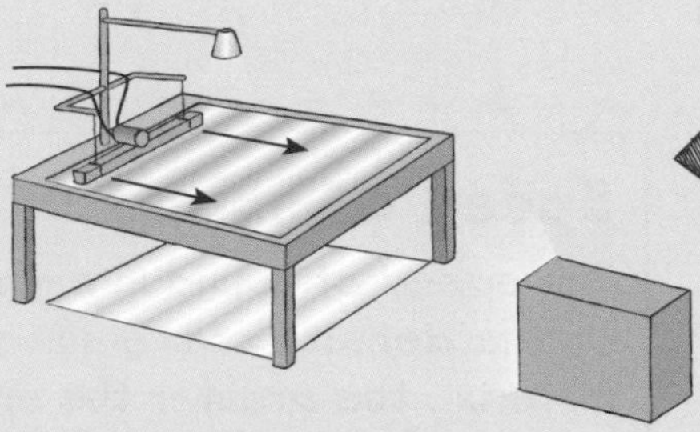

2 Calculate the frequency of the waves by counting the number of waves that pass a point each second. Do this for a minute and then divide by 60 to get a more accurate value for the frequency of the water waves.

3 Use a stroboscope to 'freeze' the waves and find their wavelength by using a ruler. The ruler can be left in the tank or the waves can be projected onto a piece of A3 paper under the tank and the wave positions marked with pencil marks on the paper.

4 Calculate the wave speed.

Results

Wavelength (m)	Frequency (Hz)	Wave speed (m/s)
0.05	10.0	5.0
0.10	5.0	5.0
0.15	3.0	4.5
0.20	2.5	5.0

Conclusion

A ripple tank can be used to determine values for the wavelength, frequency and wave speed of water waves. It is a suitable method, provided that small wavelengths and frequencies are used.

See pages 23–25 for the properties of waves, including the relationship between wave speed, frequency and wavelength.

Water and electricity are being used here, both of which can be dangerous. Be careful to take this into consideration when planning your practical.

To obtain accurate results for the wavelength of the water wave, it is best to find the distance between a large number of waves and then divide this value by the number of waves. This will reduce the percentage error in your value when determining a value for the wavelength of the wave.

Waves in water

Water waves will travel at a constant speed in a ripple tank when generated at different frequencies if the depth of the water is constant at all points. This means that the equation wave speed = frequency × wavelength will give the same wave speed – if the frequency increases, then the wavelength will decrease in proportion.

Now try this

1 Describe the main errors in determining the speed of water waves in this investigation. **(3 marks)**

2 Explain how the investigation described here can be improved. **(4 marks)**

Extended response – Waves

There will be one or more 6 mark questions on your exam paper. For these questions, you will need to think scientifically and structure your answer logically, showing how the points you make are related to each other. You can revise the topics for this question, which is about **waves**, on pages 23–30.

Worked example

Waves transfer energy without there being any overall transfer of matter. These waves can be modelled using a slinky spring, as shown in Figure 1 and Figure 2.

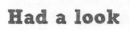

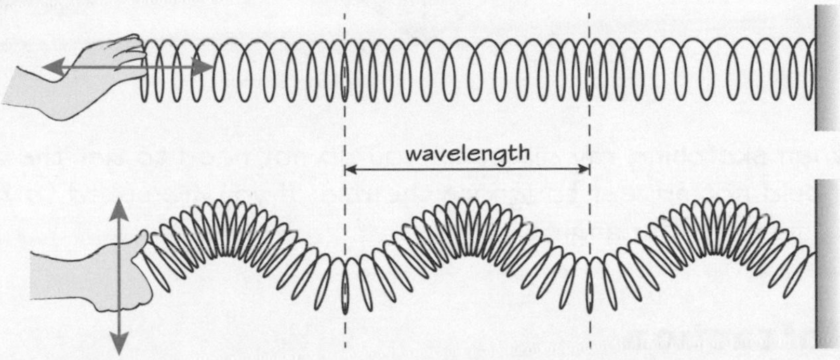

Explain how the wave motion that occurs is related to the disturbance that is causing it and the material that it is moving in.

Your answer should refer to real-life examples of both types of waves. **(6 marks)**

Figure 1 shows a longitudinal wave. A longitudinal wave is produced when the vibration is parallel to the direction of energy transfer.

Figure 2 shows a transverse wave. A transverse wave is produced when the vibration or oscillation is at right angles to the direction of energy transfer.

Sound waves, infrasound and ultrasound waves are all examples of longitudinal waves, as are seismic P waves. These waves can only transfer energy if there is a medium (a solid, liquid or gas) for them to travel through. Some transverse waves also travel through matter, with examples including water waves and seismic S waves.

Unlike longitudinal waves, some transverse waves can travel through a vacuum and do not need a medium. This is the case for all electromagnetic waves.

The wave in Figure 1 has correctly been identified as a longitudinal wave. A simple description of how it is produced is also given.

The wave in Figure 2 has correctly been identified as a transverse wave. There is also a simple description of how it is produced.

Command word: Explain

When you are asked to **explain** something, it is not enough to just state or describe something. Your answer **must** contain some reasoning or justification of the points you make.

After identifying the waves correctly, examples are given of these waves and the fact that longitudinal waves require a medium. Transverse waves can also do this, but one distinct difference is that transverse electromagnetic waves can travel through space.

All electromagnetic waves are transverse and all sound waves are longitudinal in nature. In terms of seismic waves, think of the letter 's' – a **s**ei**s**mic **S**-wave is tran**s**ver**s**e. These **S** waves will only travel through **s**olids. Unlike P waves, they will **not** travel through liquids such as the Earth's liquid outer core.

Now try this

The human ear can detect sound waves between 20 Hz and 20 000 Hz. Describe uses of the frequencies of sound waves that cannot be detected by the human ear. **(6 marks)**

Had a look ☐ Nearly there ☐ Nailed it! ☐

Reflection and refraction

Reflection and **refraction** are two properties of all types of waves.

Reflection

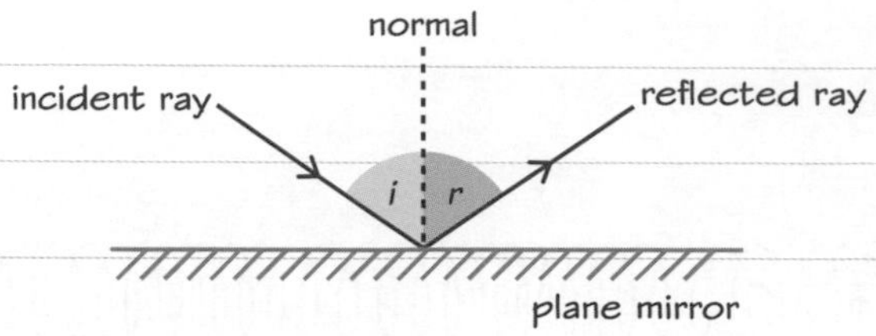

The ***normal*** is a line drawn at 90° to the mirror surface at the point that the arriving or incident ray meets the mirror. The angles ***i*** and ***r*** are measured from this line.

The law of reflection is angle ***i*** = angle ***r***

When sketching ray diagrams you do not need to get the angles exactly right, but the sketch should not appear to ignore the rule. If you are asked to *plot* a ray diagram, use a protractor to measure the angles.

Refraction

Refraction is the change in the direction of a light ray that happens when it travels from one transparent material into another. Notice that the ray of light bends *towards* the normal as it enters the glass and *away* from the normal when it leaves the glass.

Rays of light that meet a surface at 90° do not bend at all but simply continue into the material without a change in direction.

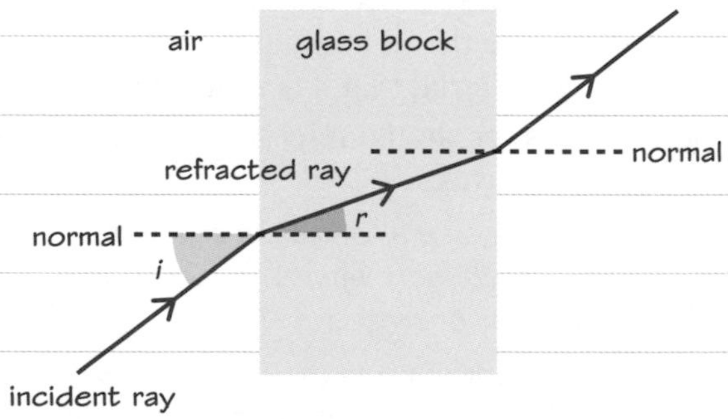

Refraction and wave speed

Refraction happens because light waves travel at different speeds in different materials. Light waves travel more slowly in glass than they do in air.

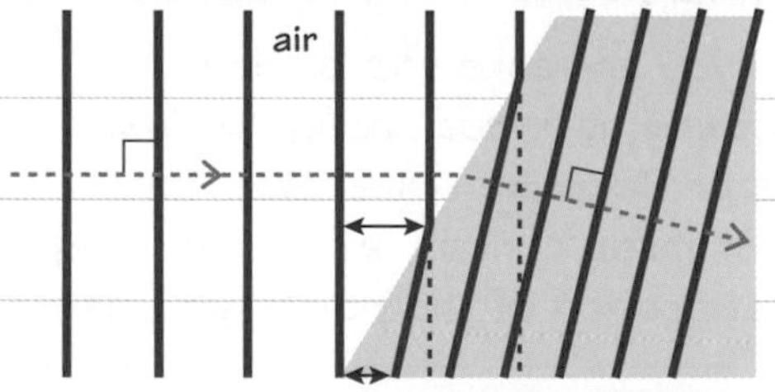

The red lines are waves and the dotted purple arrows show the direction that the light rays are travelling. When the waves travel from air into a denser material like glass they travel more slowly. The dotted red lines show how far the waves would have travelled in air but as they travel at a lower speed in glass the direction is turned through an angle. The size of this angle depends on how much the wave has been slowed down. Wavelength also changes.

Worked example

(a) State the law of reflection. **(1 mark)**

The angle of incidence is equal to the angle of reflection.

(b) Explain what happens during refraction. **(3 marks)**

The speed of the wave changes as it moves from one medium to another. Its direction also changes unless it is travelling along the normal.

Now try this

1. A ray of light strikes a plane mirror at 42° to the normal. State the angle of reflection. **(1 mark)**
2. A ray of light is travelling through a glass block before entering into air. The angle of incidence in the glass block is 34°. Estimate the angle of refraction once it has entered the air and give a reason. **(2 marks)**
3. State the factors that influence the angle of refraction for a material. **(2 marks)**
4. Describe how each of these changes during refraction:
   (a) wave speed **(2 marks)**
   (b) wavelength **(2 marks)**
   (c) frequency. **(1 mark)**

Total internal reflection

Total internal reflection involves both **reflection** and **refraction**.

The role of refraction

Refraction is when light slows down and usually changes direction when it travels from a less dense to a more dense medium.

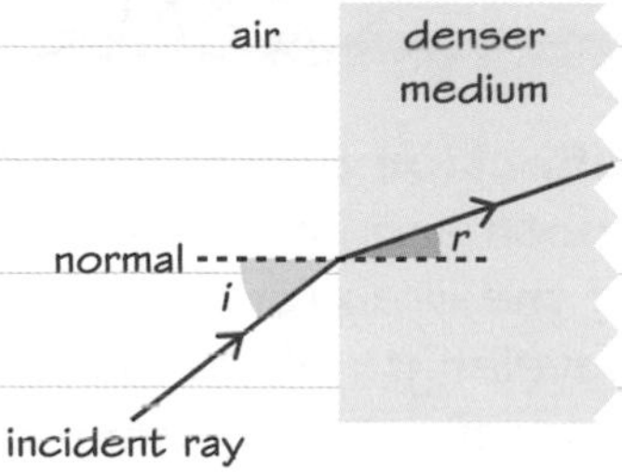

Light slows down and bends towards the normal when moving from air into glass or water.

The role of reflection

As the angle of incidence increases, the angle of refraction will increase until it reaches 90°. At this point, the ray of light is travelling along the outer surface of the glass. The angle of incidence is the **critical angle**.

Above the **critical angle**, the light is totally internally reflected.

Critical angles

Total internal reflection can only occur when light travelling from a **dense** material like glass meets a boundary with a **less dense** material like air.

Light speeds up and changes direction away from the normal when it travels from glass into air.

Total internal reflection can take place with sound as well as light.

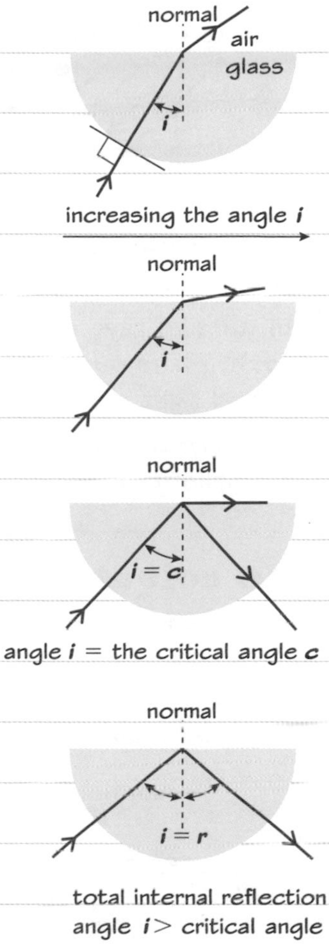

Worked example

(a) Draw a labelled diagram to show how total internal reflection (TIR) is used in optical fibres. **(3 marks)**

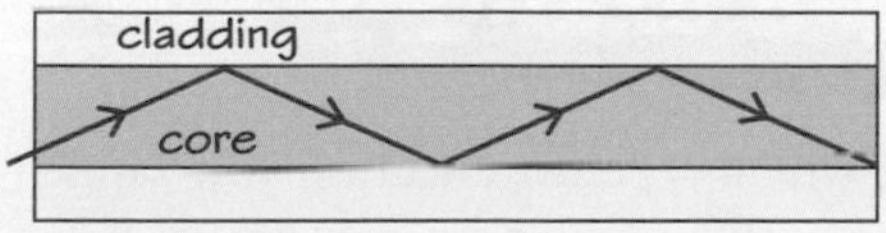

Rays of light meeting the boundary between the cladding and the core are totally internally reflected.

(b) Explain the two conditions necessary for total internal reflection to take place. **(2 marks)**

1. The cladding must have a smaller refractive index (lower density) than the core.
2. The waves must meet the boundary at an angle greater than the critical angle.

Endoscopes

Endoscopes can be used to look inside patients' bodies, and make use of optical fibres. Endoscopes allow 'keyhole' surgery. This is surgery conducted through a very small cut in the body to speed up recovery time.

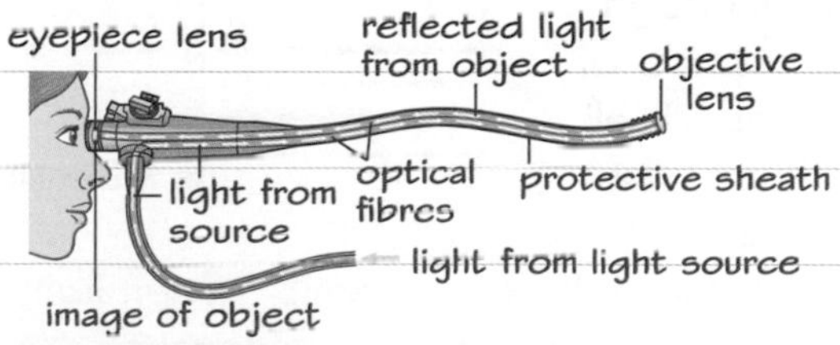

Now try this

1 Describe the conditions for total internal reflection to occur for a ray of light. **(3 marks)**
2 Describe how an endoscope works. **(3 marks)**
3 Explain the role played by reflection and refraction in an endoscope. **(4 marks)**

Colour of an object

The colour an object appears is related to the **transmission**, **reflection** and **absorption** of different wavelengths of light.

Specular reflection

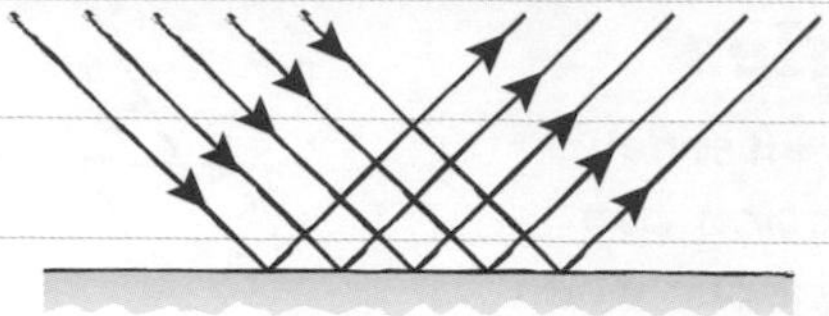

Specular reflection occurs when waves are reflected from a smooth surface.

When parallel rays of light are incident on a **smooth**, plane surface such as a mirror, the reflected light rays will also be **parallel**.

The sizes of any irregularities on the surface are much smaller than the wavelength of the wave.

Diffuse reflection

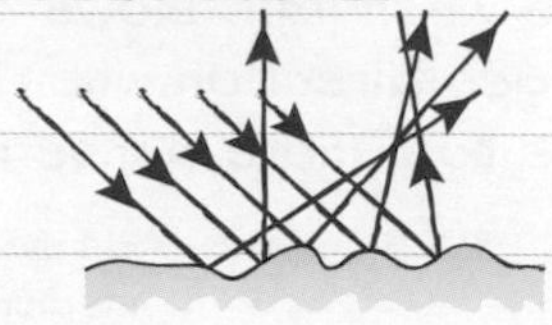

Diffuse reflection occurs when the surface is **not smooth** and has rough irregularities. The size of the irregularities is comparable with the wavelength of the wave. The incident wave is then reflected at **many different angles** and the reflected rays will not be parallel, such as when light is reflected off a painted wall.

The colour spectrum

Visible light makes up a very small part of the electromagnetic spectrum. The colours that we see can be split into different colours by a prism.

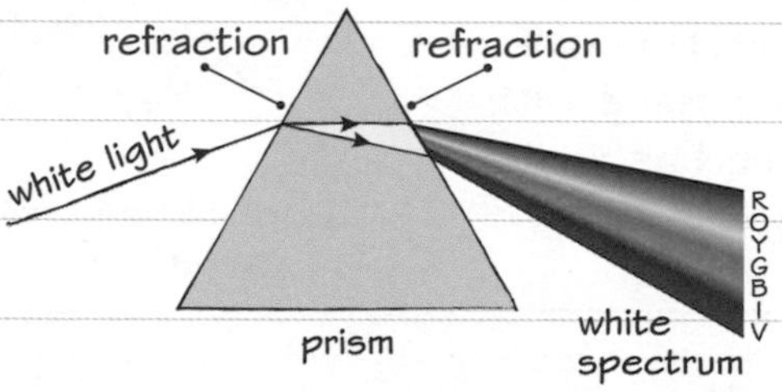

These colours all have a different wavelength, ranging from the longest wavelength at the red end of the spectrum to the shortest wavelength at the violet end.

Differential absorption at surfaces

The colour an object appears is based on how the atoms at its surface respond to the light being shone on them. A material appears green because its atoms reflect the green wavelengths and absorb all of the others.

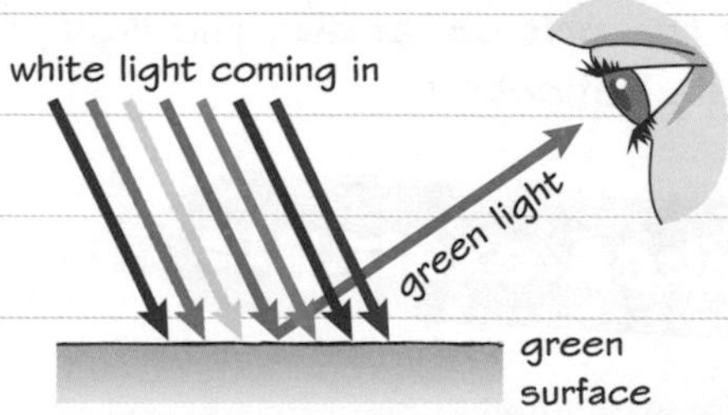

Filters

Filters let through different colours of light and absorb all other colours. For example, a green filter will let through or 'transmit' green light and absorb all of the other wavelengths.

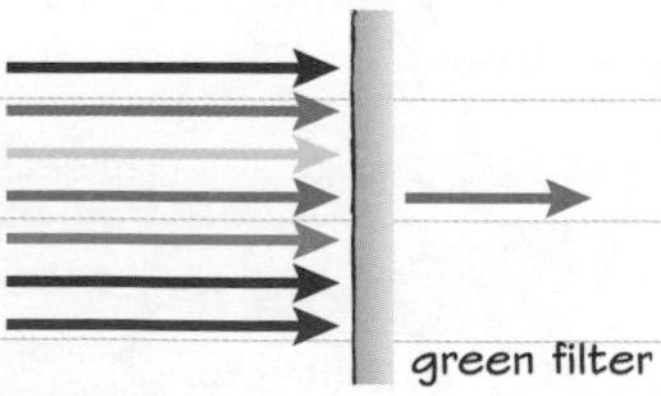

Worked example

Explain which type of reflection you would associate with a plane mirror or a calm lake surface. **(2 marks)**

specular reflection, because the surface is smooth, so parallel rays or wavefronts that are incident on these surfaces will be reflected as parallel rays or wavefronts

Now try this

1 Explain which type of reflection you would associate with a gravel path. **(2 marks)**

2 Explain how a filter works to let through red light. **(2 marks)**

3 Explain what you would see when white light travels through a red filter and then a green filter. **(2 marks)**

Lenses and power

Lenses are pieces of **glass** that **bend light** in order to bring it to a **focus**. Lenses make use of **refraction** to bend light. The more refraction that occurs, the more **power** the lens has.

Lenses and refraction

Lenses use refraction to bend light. There are two main types of lens – **a converging** lens and a **diverging** lens.

Lenses that are thicker in the middle are **converging** lenses and those that are thinner in the middle are **diverging** lenses.

converging lens

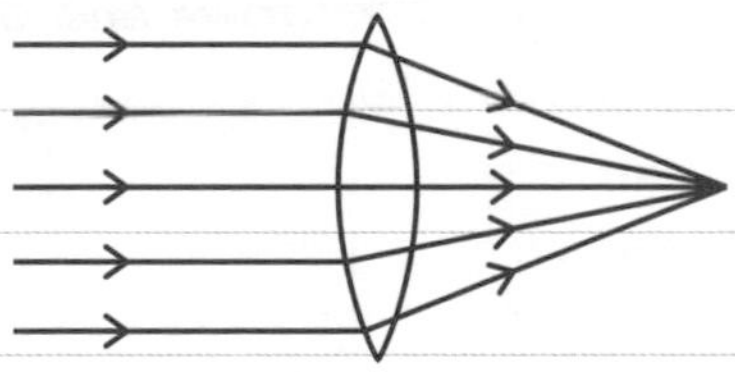

A converging lens bends rays of light towards one another, bringing them to a point.

diverging lens

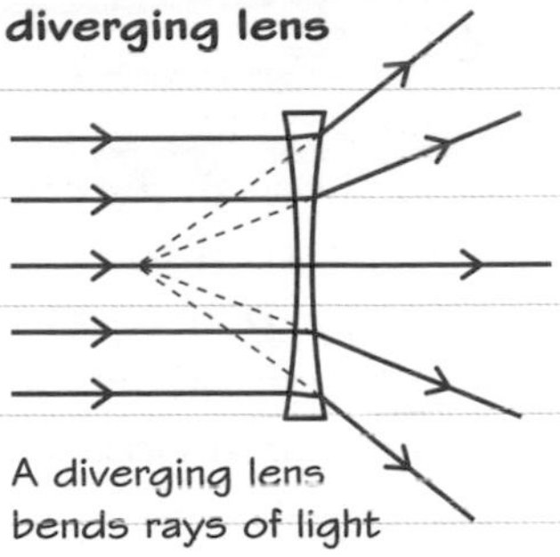

A diverging lens bends rays of light away from each other.

Power of a lens

The greater the **power** of a lens, the more it bends light. This is related to the **shape** of the lens.

Parallel rays of light all converge to pass through f.

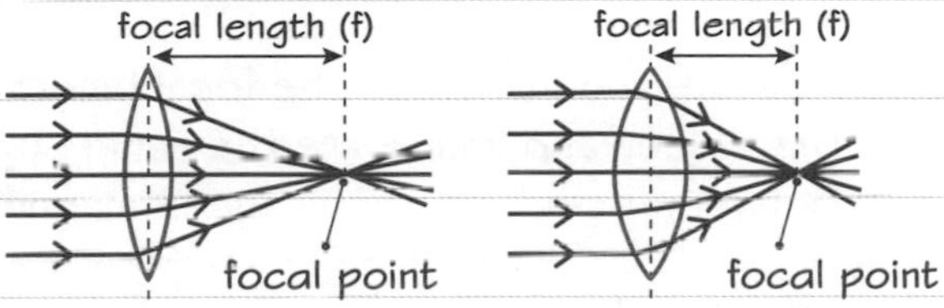

The stronger lens on the right is fatter so it has more sharply curved faces and has a shorter focal length.

Parallel rays of light diverge so that they all seem to come from f.

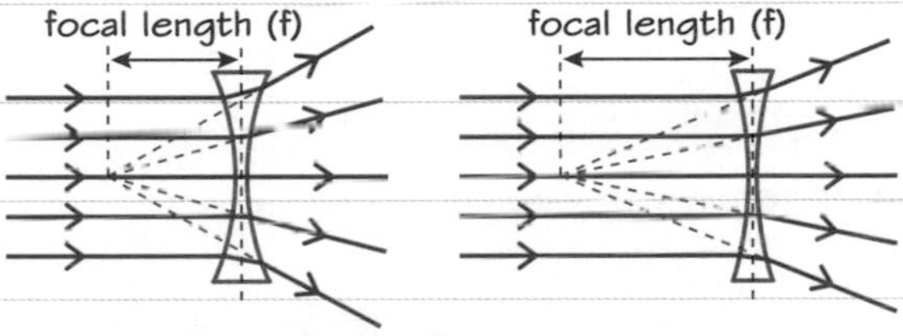

The weaker lens on the right is thinner so it has less curved faces and has a longer focal length.

Worked example

Describe how the following lenses compare in terms of their (a) powers and (b) focal lengths. **(2 marks)**

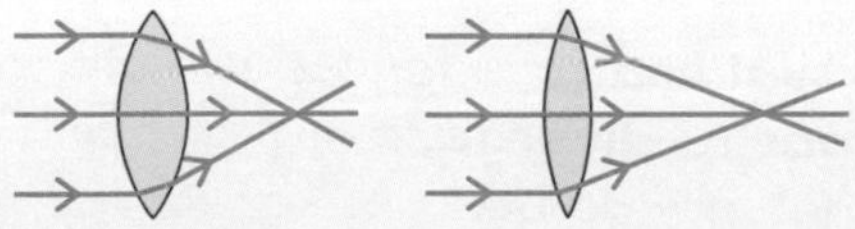

lens a   lens b

(a) Lens a is more powerful than lens b.

(b) Lens b has a longer focal length than lens a.

For lenses made of the same type of glass, the thicker, more curved lens will have the greater power and shorter focal length.

Now try this

1 Explain how the power of a lens is related to its shape. **(2 marks)**

2 What is the relationship between the power of a lens and its focal length? **(1 mark)**

3 For a long-sighted person, light rays from a point near to the eye are brought to a focus at a point beyond the retina. Suggest how the person's vision could be corrected. **(2 marks)**

Real and virtual images

The images produced by a lens depend on the **type of lens** being used and the **position of the object** with respect to the lens.

Real images and converging lenses

When parallel light rays from a distant object pass through a converging lens, they are brought to a focus at a point – the **focal point** or **principal focus**. The object can be said to be 'at infinity' if its distance from the lens is much greater than the focal length of the lens.

A **real image** is an image that can be **produced on a screen**. The lens focuses light rays at the screen.

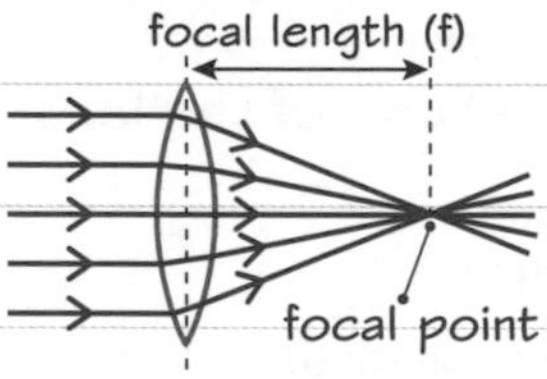

The focal length depends on the **thickness** of the lens and the **material** that the lens is made from.

A **real image** is formed where light rays **converge** and are **actually focused** on a screen.

The nature of a real image

Far from the lens, the real image is upside down and smaller than the object.

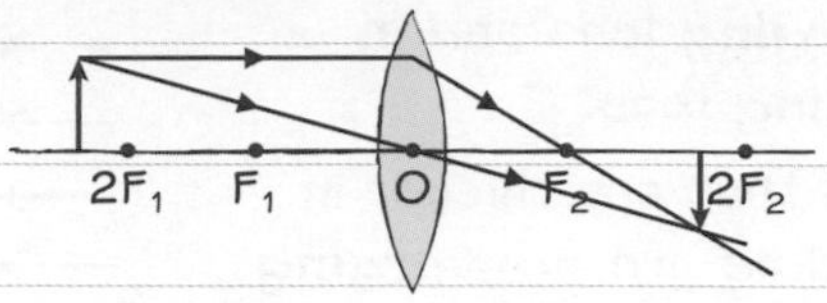

Moving the object closer to the lens causes the image to become larger. The position of the image will move away from the lens but will remain real and upside down.

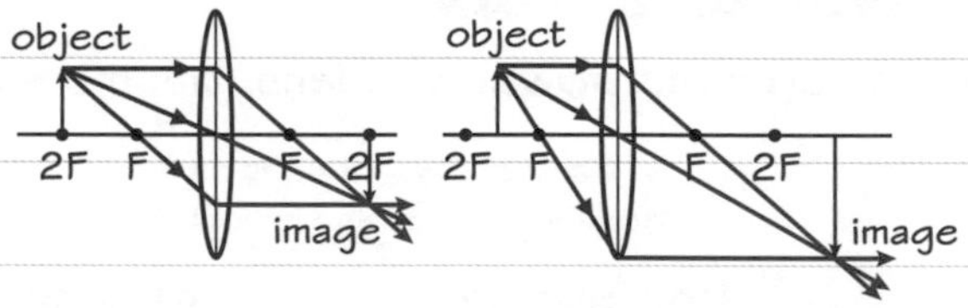

At a distance of twice the focal length (2F), the object and image are the same size. Between F and 2F, the image is magnified.

Virtual images, converging and diverging lenses

A **virtual image** is formed by a converging lens when the object is between the focal point and the lens.

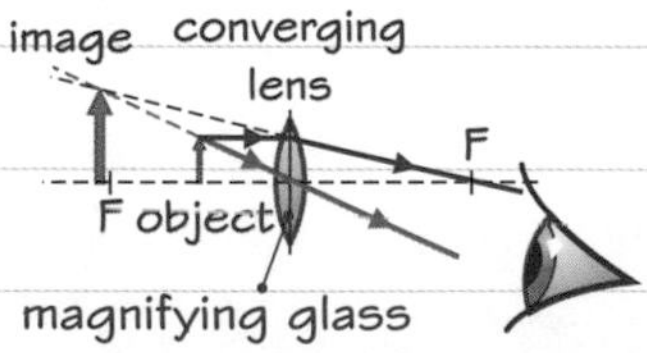

A diverging lens produces a virtual image.

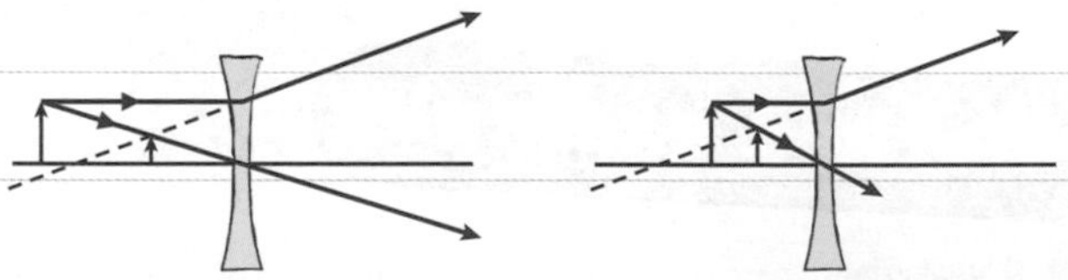

object outside focal point | object inside focal point

A **virtual image** is formed by light rays which **appear** to **diverge** from that point, but do not actually do so.

Worked example

State what type of image is formed by:

(a) the human eye **(1 mark)**

a real image

(b) a magnifying glass. **(1 mark)**

a virtual image

Now try this

1 State what type of image is produced on a screen at the cinema. **(1 mark)**

2 Explain why real images cannot be produced by converging lenses when the object being viewed is closer to the lens than a distance equal to the focal length of the lens. **(3 marks)**

Electromagnetic spectrum

Infrared radiation, **visible light** and **ultraviolet radiation** are all part of the **electromagnetic spectrum.**

All electromagnetic waves...

- are **transverse waves** (the electromagnetic vibrations are at right angles to the direction the wave is travelling – see page 23)
- travel at the **same speed** (3×10^8 m/s) in a **vacuum**
- **transfer energy** to the observer.

Watch out!

The different parts of the electromagnetic spectrum have different properties, which you will read about on the following pages. But it is important to remember that some of their properties are *the same*. They are *all* transverse waves, and *all* travel at the same speed in a vacuum.

The electromagnetic spectrum

As the **frequency** of the radiation **increases**, the **wavelength decreases**.

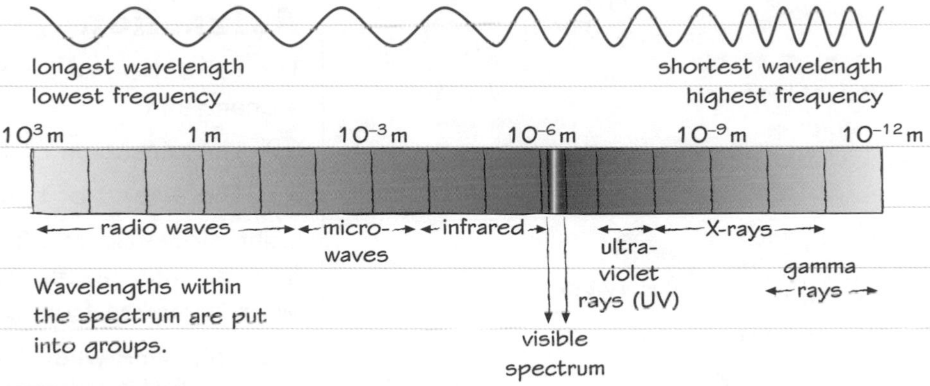

Worked example

(a) List the main groups of electromagnetic waves from longest to shortest wavelength. **(1 mark)**

radio, microwave, infrared, visible, ultraviolet, X-rays, gamma

You can use a mnemonic to help you to remember the order. A mnemonic is a sentence or phrase whose words have the same initial letters as the list you are trying to remember. For example: **R**ed **M**onkeys **I**n **V**ans **U**se **X**-ray **G**lasses.

(b) Calculate the frequency of an electromagnetic wave travelling in air that has a wavelength of 5×10^{-7} m. **(4 marks)**

frequency = wave speed ÷ wavelength

frequency = $3 \times 10^8 \div 5 \times 10^{-7}$

$= 6 \times 10^{14}$ Hz

You can find the wave equation on page 24.

Now try this

1 State the type of electromagnetic radiation that has the (a) longest wavelength, (b) highest frequency. **(2 marks)**

2 The human eye detects wavelengths that range from 4×10^{-7} m to 7×10^{-7} m. Work out the range of frequencies that the eye detects. **(3 marks)**

3 An electromagnetic source produces 4×10^{18} waves in 1 minute.
 (a) Calculate the wavelength of the source. **(3 marks)**
 (b) State what part of the electromagnetic spectrum this radiation belongs to. **(1 mark)**

Investigating refraction

Practical skills You can investigate refraction in rectangular glass blocks in terms of the interaction of light waves with matter.

Core practical

Aim

To investigate the nature of how light waves change direction when they move from air into glass.

Apparatus

ray box and slits, 12 V power supply, glass block, protractor, A3 paper, sharp pencil

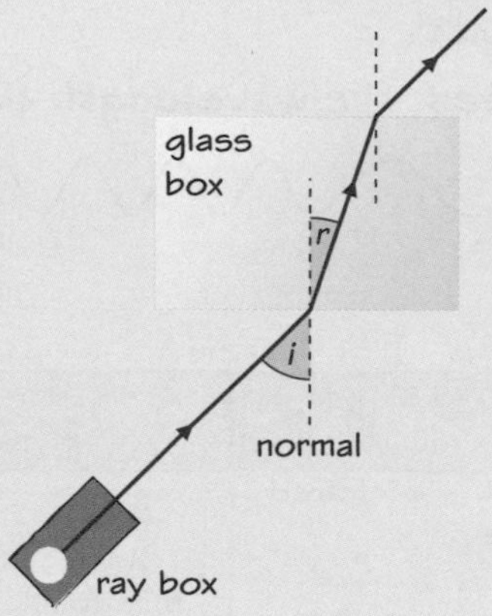

Method

1 Place the rectangular block on the A3 paper and draw around it with the sharp pencil.
2 Draw the normal line, which will be at right angles to the side of the block towards which the light ray will be shone.
3 Using the protractor and pencil, mark on the paper angles of incidence of 0° through to 80° in 10° intervals.
4 Starting with the 0° angle (the light ray travelling along the normal line), direct the light ray towards the block and mark its exit point from the block with a sharp pencil dot.
5 Remove the glass block and join the dot to the point of incidence by drawing a straight line. Measure and record this angle, which is the angle of refraction.
6 Repeat for all of the other angles from 10° to 80°.

Results

Angle of incidence	Angle of refraction
0°	0°
10°	7°
20°	13°
30°	19°
40°	25°
50°	31°
60°	35°

Conclusion

When a light ray travels from air into a glass block, its direction changes and the angle of refraction will be less than the angle of incidence unless it is travelling along the normal.

There is more information about the refraction of light on pages 32 and 33.

When carrying out the investigation, make sure that you direct a thin beam of light towards the point at which the normal makes contact with the glass block. Clearly mark the angles from 0° to 80° with a sharp pencil and ruler.

Refraction

The refraction of a light ray involves a change in:

- ✓ the direction of the light ray
- ✓ the speed of the light.

Light slows down when it moves from air into glass and speeds up when it moves from glass into air.

The only time when the direction does not change is when the beam is travelling along the normal.

Remember that the angle of incidence is measured with respect to the normal line.

Light will slow down more when it travels from air into glass than it will when it travels from air into water. This is because the 'optical density' of glass is greater than that of air.

Now try this

1 State the two things that can change when a light ray is refracted. **(2 marks)**

2 Draw a table of results for a light ray travelling from air into a block of water instead of a block of glass. **(2 marks)**

3 Explain how the (a) frequency and (b) wavelength of a light ray are affected when it enters glass from air. **(4 marks)**

Wave behaviour

The behaviour of **electromagnetic waves (EM)** in different materials depends on their **wavelength** and **velocity**.

Electromagnetic waves

All electromagnetic waves travel at the **same speed in a vacuum**. This is the **speed of light**, 3×10^8 m/s. Our eyes detect the part of the electromagnetic spectrum that is visible light, but there are other wavelengths that behave differently.

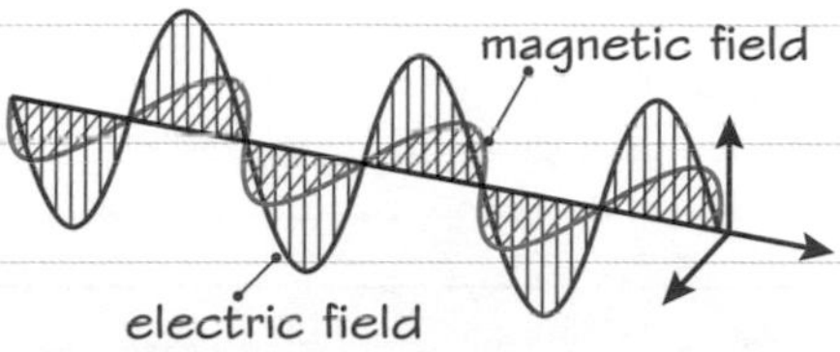
magnetic field
electric field

Electromagnetic waves are transverse. They are composed of an electric field and a magnetic field at right angles to one another.

Behaviour of EM waves

Electromagnetic waves may be:

1. reflected off a surface
2. refracted when they move from one material to another
3. transmitted when they pass through a material
4. absorbed by different materials. For example, UV is absorbed by the skin but not by the Earth's atmosphere.

The extent to which these four things happen depends on the material and the wavelength of the EM waves.

Radio waves and microwaves

Radio waves and microwaves are both used for communicating. They have different wavelengths and behave differently in different materials.

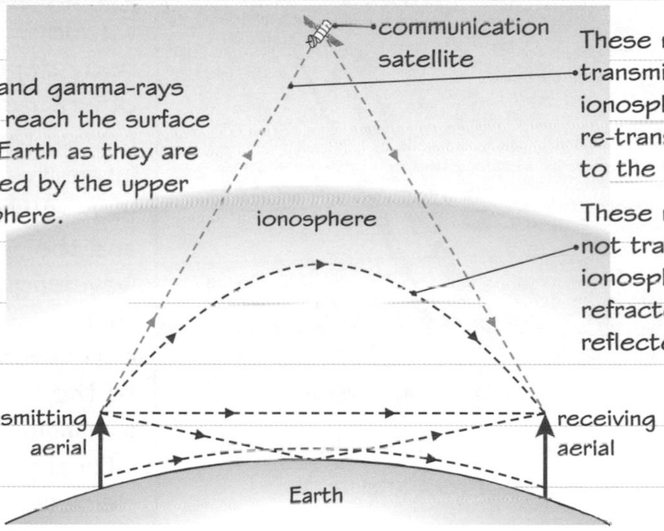
communication satellite
X-rays and gamma-rays cannot reach the surface of the Earth as they are absorbed by the upper atmosphere.
These microwaves are transmitted by the ionosphere and then re-transmitted back to the receiver.
ionosphere
These radio waves are not transmitted by the ionosphere; they are refracted and then reflected to the receiver.
transmitting aerial
receiving aerial
Earth

Worked example

Explain why it is not possible to get a suntan if you are inside a house, but it is possible to listen to the radio. **(4 marks)**

The ultraviolet rays that cause suntans are not transmitted by the walls of houses, they are absorbed by them. Radio waves are transmitted by the walls of houses so can be detected by a radio receiver inside the house.

Wave velocities

Electromagnetic waves have different velocities in different materials. This is linked, to an extent, to the density of the material. The speed of light in a diamond is around 40% of the speed of light in a vacuum or air, due to the high optical density of diamond.

Now try this

1 State four things that can happen to electromagnetic waves. **(2 marks)**

2 Explain why we do not need a communication satellite to relay some radio waves. **(2 marks)**

Temperature and radiation

All objects above absolute zero will **emit** electromagnetic radiation. Their temperature will depend on how much of it they absorb and how much of it they radiate.

Emitting radiation

A hot cup of tea at 90°C will emit radiation that is mainly in the **infrared part** of the EM spectrum, whereas the Sun's surface temperature of 5700°C means that it emits **visible light** and **ultraviolet** radiation which have a shorter wavelength than infrared. Bodies that are much hotter than the Sun will emit **X-rays**.

Factors affecting temperature

1. A body at a constant temperature absorbs the same amount of radiation as it emits.
2. An object will increase its temperature if it absorbs more radiation than it emits.
3. An object will decrease its temperature if it emits more radiation than it absorbs.

As the temperature of a body increases, more energetic radiation is emitted, including infrared, visible and UV.

Factors affecting the temperature of the Earth

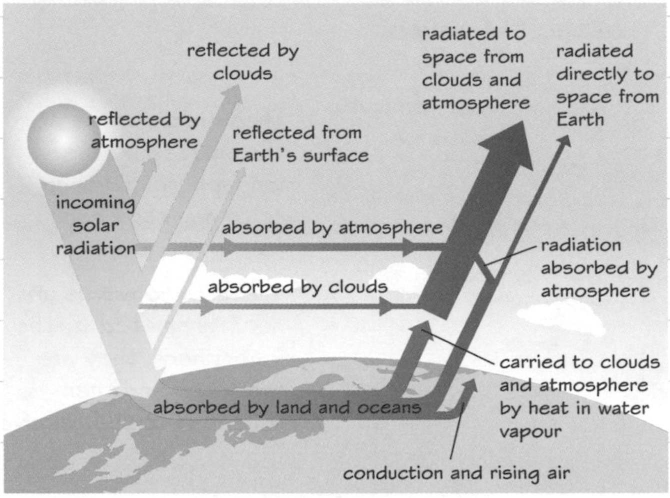

The temperature of the earth depends on how much radiation is absorbed and how much is emitted.

Intensity and wavelength

Intensity is the energy emitted per square metre per second – or the power emitted per square metre. The intensity and wavelength of the radiation emitted by a body depend on its temperature. As the object gets hotter, the intensity increases and the wavelength that corresponds to the maximum intensity decreases.

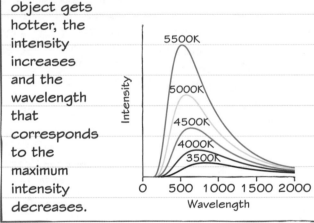

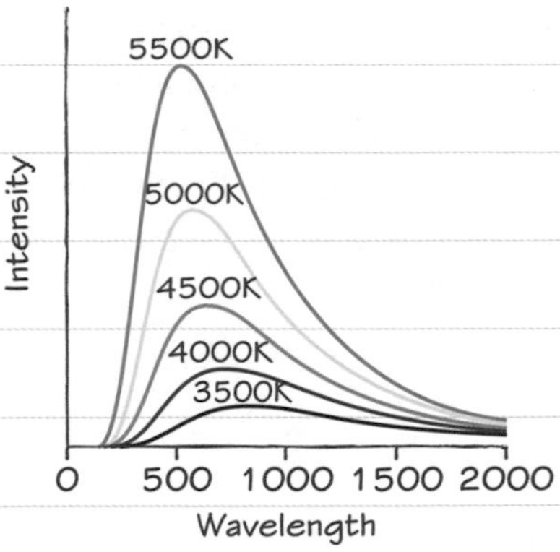

Worked example

Explain why the temperature of the Earth changes over the course of the day. **(2 marks)**

During the day, the side of the Earth facing the Sun absorbs more radiation than it emits and so it warms up. The opposite happens at night when facing away from the Sun.

The temperature of the Earth depends on the balance between the incoming radiation and the radiation emitted. Both will happen at the same time, but the relative amount of each determines whether the Earth's temperature goes up or down.

Now try this

1 Describe what you know about a body when its temperature is increasing. **(2 marks)**

2 Explain the conditions needed for the Earth to remain at a constant temperature. **(2 marks)**

3 Explain why the colour of a star depends on its surface temperature. **(4 marks)**

Thermal energy and surfaces

Practical skills You can investigate the relationship between the **rate** at which a material **emits** or **absorbs thermal radiation** and the nature of its **surface**.

Core practical

Aim

To investigate how the nature of a surface affects the rate at which it absorbs or radiates thermal energy.

Apparatus

Leslie's cube, thermometers, stopwatch, ruler, clamp, boss and retort stand.

Method 1: radiating thermal energy

1 Fill Leslie's cube with hot water at a known temperature. Place a thermometer a small distance away from each of the sides. Plug the top of the cube with a bung.

2 Measure the temperature at a distance of 10 cm from each of the four sides of Leslie's cube for a period of 5 minutes, taking a reading every 30 seconds.

Method 2: absorbing thermal energy

1 Fill Leslie's cube with the same volume of cold water at a known temperature. Plug the top of the cube with a thermometer and bung.

2 Heat each side, one at a time, with a radiant heater from the same distance away (about 10 cm).

3 Record how the temperature of the water changes, every 30 seconds, over an appropriate period of time.

Results

Method 1: Plot graphs of temperature against time for each of the four surfaces on the same axes.

	Dull black	Shiny black	Shiny white	Dull white
Start	80 °C	80 °C	80 °C	80 °C
Finish	56 °C	62 °C	74 °C	68 °C

Method 2: Plot a graph of temperature against time.

	Dull black	Shiny black	Shiny white	Dull white
Start	20 °C	20 °C	20 °C	20 °C
Finish	34 °C	28 °C	22 °C	24 °C

Conclusion

Typical results obtained are shown in the table.

	Dull black	Shiny black	Shiny white	Dull white
Emitter	Best	2nd	3rd	Worst
Absorber	Best	2nd	3rd	Worst

There is more information on the emission and absorption of thermal energy on pages 39 and 40.

Be careful when using hot water as it may scald.

Conduct the investigation accurately and fairly by ensuring that the thermometer used is the same for each surface, the distance from the surface is the same and the starting temperature is the same in each case. Take the same number of readings over the same time period and intervals.

Leslie's cube

Leslie's cube is commonly used in schools to demonstrate how the nature of a surface affects the rate at which thermal energy is absorbed or emitted.

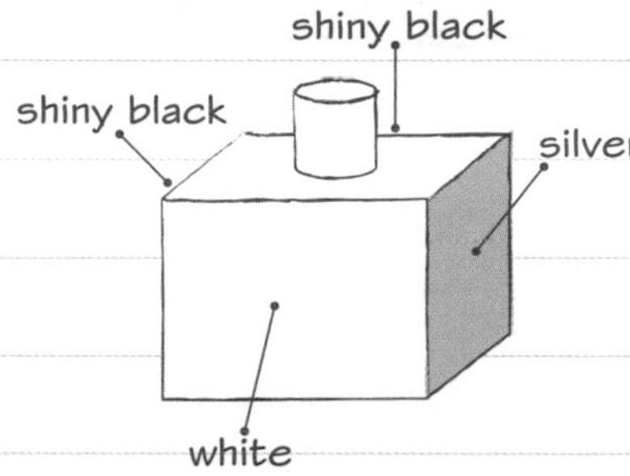

Four of its sides have different natures, as shown.

Now try this

1 Sketch four curves on a temperature–time graph for the four sides of a Leslie's cube that has been filled with hot water. **(4 marks)**

2 Explain why the starting temperature for the Leslie's cube practical needs to be kept constant. **(4 marks)**

Dangers and uses

The amount of **energy** that is transferred by an electromagnetic wave is dependent on its **wavelength** or **frequency**. The **highest frequencies** (shortest wavelengths) are the most energetic and the **most dangerous** waves.

Dangers

X-rays and gamma-rays can cause **mutations** (changes) to the DNA in cells in the body. This may kill the cells or cause **cancer**.

UV in sunlight can damage skin cells, causing **sunburn**. Over time, exposure to UV can cause **skin cancer**. UV can also damage the **eyes** leading to eye conditions.

Infrared radiation transfers thermal energy. Too much infrared radiation can cause **skin** burns.

Microwaves heat water – so they can heat the water inside our bodies. Heating cells can damage or kill them.

high frequency

gamma-rays

X-rays

ultraviolet

visible light

infrared

microwaves

radio waves

low frequency

Uses

- to sterilise food and medical equipment
- in scanners to detect cancer
- to treat cancer

- to look inside objects, including medical x-rays to look inside bodies
- in airport security scanners, to see what people have in their luggage

- to detect security marks made using special pens
- inside fluorescent lamps
- to detect forged banknotes (real banknotes have markings that glow in UV light)
- to disinfect water

- allows us to see, lights up rooms and streets, buildings and roads (illumination)
- photography

- in cooking (by grills and toasters)
- to make thermal images (images using heat), used by police and rescue services
- in short-range communications, such as between laptops or other small computers
- in remote controls for TVs and other appliances, where the signal only has to travel short distances
- to send information along optical fibres
- in security systems such as burglar alarms, to detect people moving around

- in mobile phones, and to communicate by satellite transmissions
- for cooking (in microwave ovens)

- broadcasting radio and TV programmes
- communicating with ships, aeroplanes and satellites

Worked example

State which type(s) of electromagnetic radiation are:

(a) used for communicating   **(4 marks)**

radio waves, microwaves, infrared and visible light

(b) used for cooking   **(2 marks)**

microwave and infrared

(c) used for cleaning things   **(2 marks)**

ultraviolet and gamma

(d) most dangerous.   **(3 marks)**

ultraviolet, X-rays and gamma

Radio waves can be produced by, or can themselves induce, oscillations in electrical circuits.

The different types of electromagnetic radiation share uses and dangers. For example, more than one type can be used for communicating and cooking, and there are three main types of electromagnetic radiation that can cause humans serious harm.

Now try this

1 Explain how X-rays are useful to doctors but can also be harmful to patients.   **(3 marks)**

2 Explain how gamma-rays can be used to detect cancer even though they can also cause it.   **(2 marks)**

3 Compare the use of microwaves and infrared radiation to cook food.   **(5 marks)**

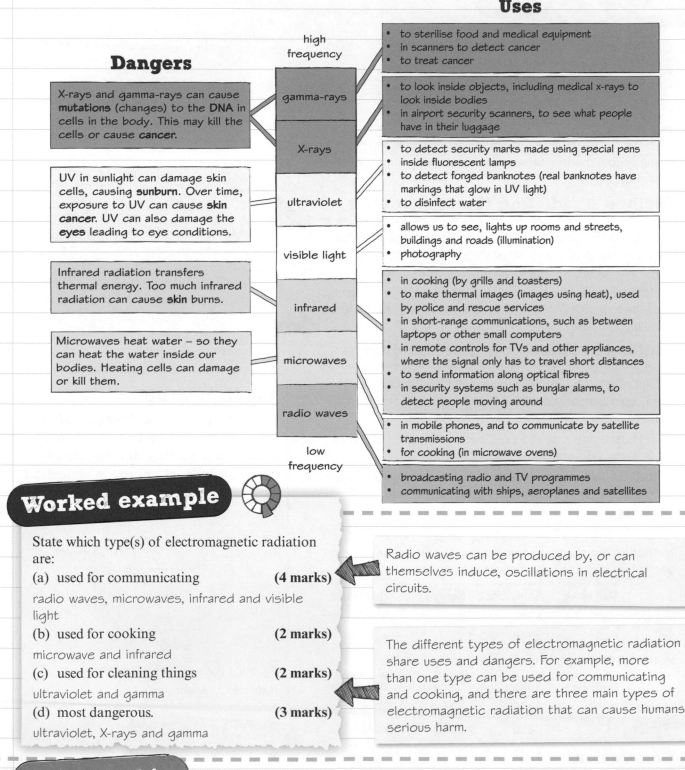

Changes and radiation

Radiation is **absorbed** or **emitted** when **electrons** jump between **energy levels**.

Energy levels

Electrons can only exist in atoms at certain, well-defined energy levels. These **energy levels** depend on the atom, and the electrons inside the atom can move between the shells or leave the atom completely.

Electromagnetic radiation is **emitted** or **absorbed** by atoms based on whether energy is given out or taken in.

Electrons move up energy levels when they absorb energy and they fall down to lower energy levels when they emit energy.

Electrons move from a lower energy level to a higher energy level when the correct amount of energy is absorbed.

Electromagnetic radiation is emitted when electrons fall down from a higher to a lower energy level.

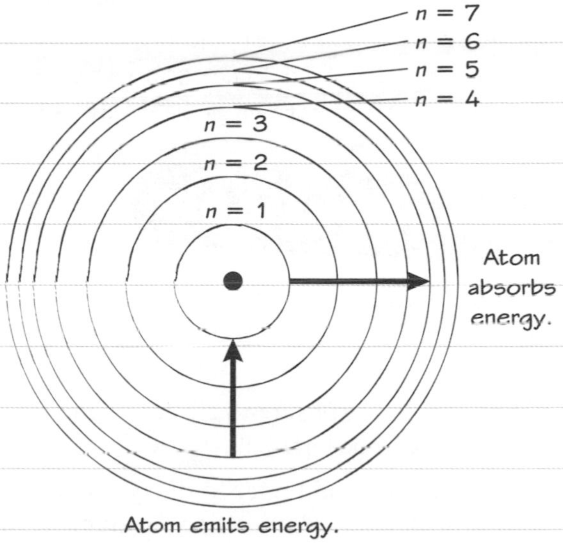

The electromagnetic radiation can have a **wide frequency range**, from **infrared** through to **ultraviolet** and beyond.

Nuclear radiation

Energy is also emitted from the **nuclei** of **unstable** atoms. **Protons** and **neutrons** also occupy **energy levels** in the nucleus, in the same way that electrons do in the atom.

When energy changes occur in the nucleus, high-energy **gamma-rays** are emitted.

Gamma-rays can be emitted over a large range of frequencies, depending on the energy levels within the nucleus.

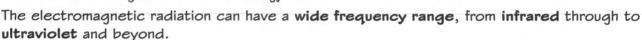

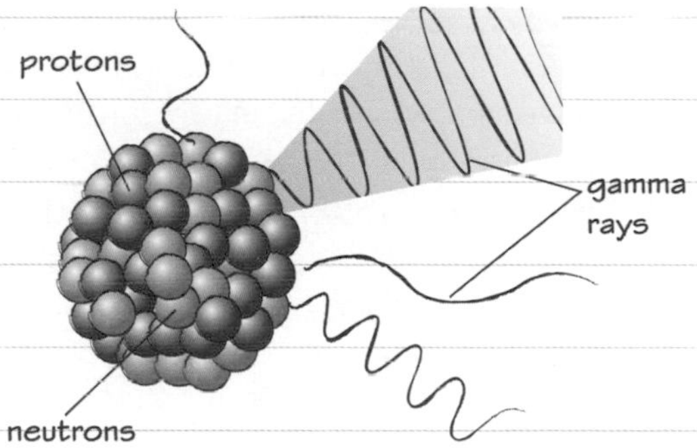

Worked example

Explain the energy changes that occur within the electron levels of the atom. **(3 marks)**

Electrons can move between energy levels if they absorb or emit electromagnetic radiation. When they absorb electromagnetic radiation they move up energy levels and when they emit electromagnetic radiation they move down energy levels.

The energy of electromagnetic radiation that is emitted or absorbed has to **match** the difference between the **energy levels** that the **electron** is moving between.

Energy changes in the **nucleus** are far greater than those seen between electron energy levels.

Now try this

1 Describe what happens when electrons
(a) absorb electromagnetic radiation **(2 marks)**
(b) emit electromagnetic radiation. **(2 marks)**

2 Explain why visible light is emitted when electrons move between shells but gamma-rays are emitted when there are changes in the nucleus. **(4 marks)**

Extended response – Light and the electromagnetic spectrum

There will be one or more 6 mark questions on your exam paper. For these questions, you will need to think scientifically and structure your answer logically, showing how the points you make are related to each other. You can revise the topics for this question, which is about **light and the electromagnetic spectrum**, on pages 32–42.

Worked example

Exposure to certain types of electromagnetic radiation can have harmful effects on the human body.

Describe how exposure to electromagnetic radiation can be harmful.

Your answer should refer to the types of electromagnetic radiation and the damage that they may cause. **(6 marks)**

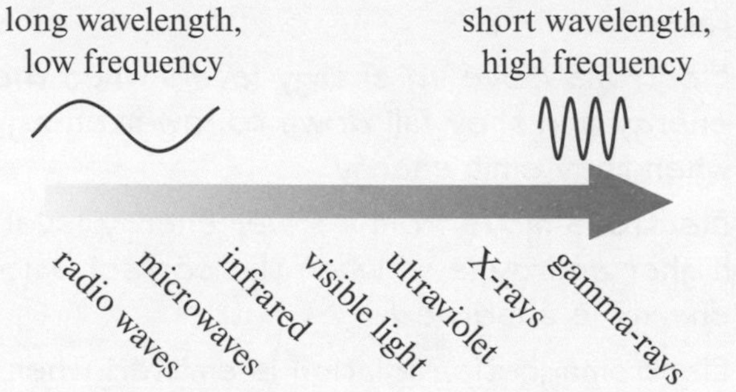

Apart from radio waves and visible light, which are not deemed to cause damage to humans, the other five members of the electromagnetic spectrum do cause harmful effects.

Microwaves are absorbed by water molecules and can lead to internal heating of body cells, which can be harmful. Infrared radiation can cause burns to the skin.

Ultraviolet radiation can damage the eyes, leading to eye conditions. It can also cause skin cancer as surface cells are affected. X-rays and gamma-rays can both lead to cellular damage and mutations, which can lead to diseases such as cancer.

The initial sentence states which types of electromagnetic wave are dangerous and which are not. This will help to structure the rest of the answer.

Two of the low-energy members of the electromagnetic spectrum are commented on next.

The three higher-energy electromagnetic waves are then commented on. The energy of electromagnetic radiation is related directly to its frequency, so as frequency increases, so does the energy and so does the extent of the damage.

UV, X-rays and gamma-rays are examples of **ionising radiation**. This can lead to tissue damage and possible mutations in cells. Although not mentioned here, you could be asked a question that relates specifically to the dangers of ionising radiation. Of the three types mentioned here, only gamma rays come from the nucleus of the atom. You can read more about this on pages 45–62.

Command word: Describe

When you are asked to **describe** something, you need to write down facts, events or processes accurately.

Now try this

Describe how the electromagnetic spectrum is useful to humans. Refer to specific members of the EM spectrum in your answer. **(6 marks)**

Structure of the atom

There is a very big difference in the sizes of **atoms**, **nuclei** and **molecules**.

Structure of the atom

Atoms have a **nucleus** containing **protons** and **neutrons**. **Electrons** move around the nucleus of an atom. An atom has the same number of protons and electrons, so the **+** and **– charges** balance and the atom has no overall charge.

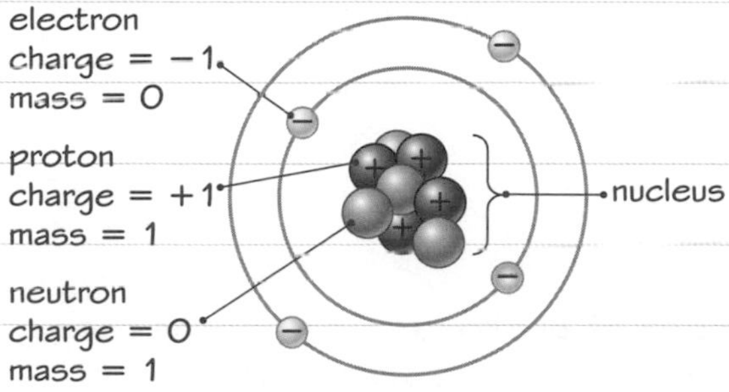

The atom and the nucleus

1. All atoms have a **nucleus**. The **nucleus** is always **positively charged** as it contains **protons** which have a positive charge and neutrons which do not have a charge.
2. The **nucleus** contains more than **99%** of the **mass** of the atom.
3. The total number of **protons** in an atom's **nucleus** must be the same as the total number of **electrons** in the shells.
4. Electrons in atoms always orbit the **nucleus** and have a **negative charge**.
5. Atoms are always **neutral** because the positive charge from the **protons** cancels out the negative charge from the **electrons**.
6. The **nucleus** of an atom of an element may contain **different numbers of neutrons**.

Atom, nucleus and molecule

A **molecule** is two or more atoms bonded together. A **molecule** is about 10 times the diameter of an atom and about 10^6 times the diameter of a nucleus.

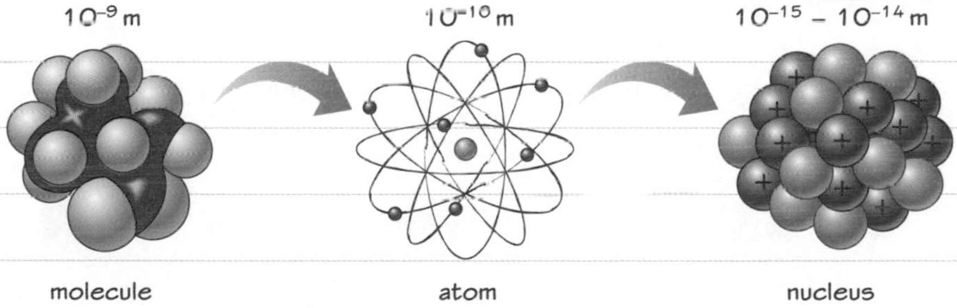

Gases such as oxygen and carbon dioxide are **molecules**. Water is also a **molecule**.

The diameter of an atom is about 10^{-10} m. The diameter of the nucleus is about 10^{-15} m.

Diagrams of atoms are always incorrect in terms of the scale. In reality, if the nucleus were the size shown here, the nearest electron would be at least 1000 m away.

Worked example

A molecule has a length of 4.8×10^{-9} m. A nucleus has a diameter of 1.2×10^{-15} m. How many nuclei would need to be placed side by side in a line to have the same length as the molecule? **(3 marks)**

4.8×10^{-9} m $\div$ 1.2×10^{-15} m
$= 4 \times 10^6$ or 4 million nuclei.

Maths skills When dividing values in standard form, remember that you need to
- divide the numbers at the front
- subtract the powers.

i.e., $A \times 10^x \div B \times 10^y = (A/B) \times 10^{(x-y)}$.

Remember, $-9 - (-15)$ is equal to 6 because two minuses make a plus, so $-9+15 = 6$.

Now try this

A poster shows a diagram of an atom that is not drawn to scale. The overall diameter of the atom on the poster is 60 cm. Calculate the diameter of the nucleus on the poster if it was properly drawn to scale. **(4 marks)**

Atoms and isotopes

Atoms of the same **element** can be **different** based on the composition of their **nuclei**.

Atoms

Atoms are made up of **protons**, **neutrons** and **electrons**. The **protons** and **neutrons** are in the nucleus of the atom, and the **electrons** move around the outside.

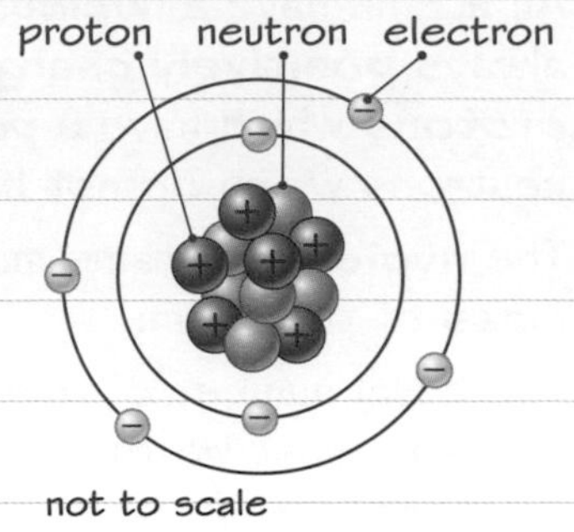

Describing atoms

All the atoms of a particular element have the same number of **protons**. The number of protons in each atom of an element is called the **atomic number**, or **proton number**.

The total number of protons and neutrons in an atom is the **mass number**, or **nucleon number**.

The atomic number and mass number of an element can be shown like this:

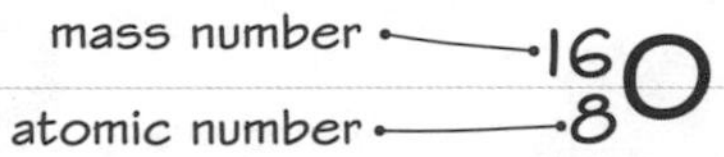

Isotopes

Atoms of a particular element always have the **same number of protons**, but they can have **different numbers of neutrons**. Atoms with the **same number of protons** but **different numbers of neutrons** are **isotopes** of the same element.

As the number of protons determines the element, you can also say that the nuclei of different elements have their own characteristic charge. In this example, a carbon nucleus always has a charge of +6.

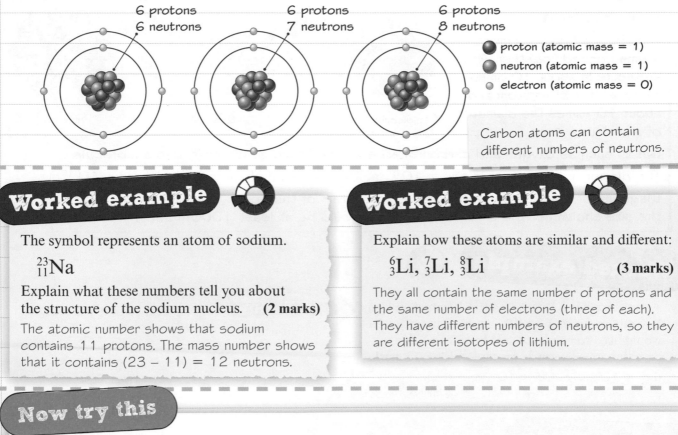

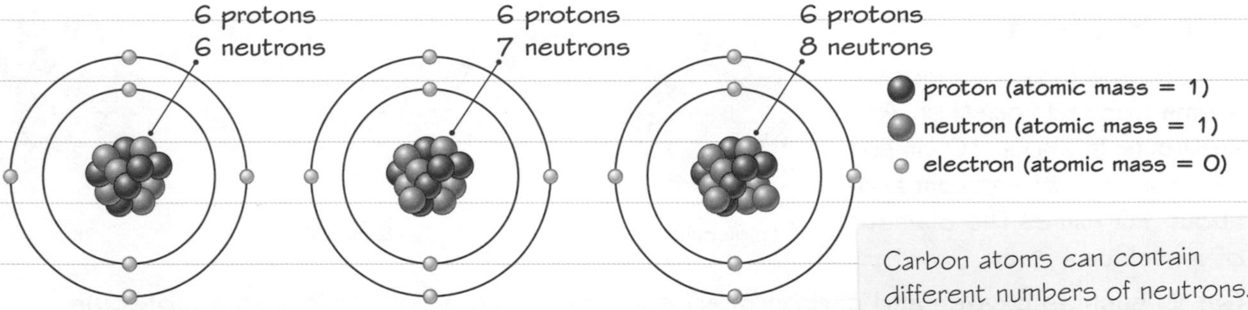

Carbon atoms can contain different numbers of neutrons.

Worked example

The symbol represents an atom of sodium.

$^{23}_{11}Na$

Explain what these numbers tell you about the structure of the sodium nucleus. **(2 marks)**

The atomic number shows that sodium contains 11 protons. The mass number shows that it contains (23 − 11) = 12 neutrons.

Worked example

Explain how these atoms are similar and different:

$^{6}_{3}Li$, $^{7}_{3}Li$, $^{8}_{3}Li$ **(3 marks)**

They all contain the same number of protons and the same number of electrons (three of each). They have different numbers of neutrons, so they are different isotopes of lithium.

Now try this

1 An atom of boron (B) has 5 protons and 6 neutrons. Show the atomic number and mass number as a symbol for the isotope. **(1 mark)**

2 Explain how these atoms are similar and different: $^{14}_{7}N$ $^{15}_{7}N$. **(3 marks)**

Atoms, electrons and ions

Electrons can **move** between energy levels, or they can **leave** the atom completely.

Electrons in orbit

In all atoms, electrons orbit the nucleus in different orbits or energy levels, at different, fixed distances from the nucleus.

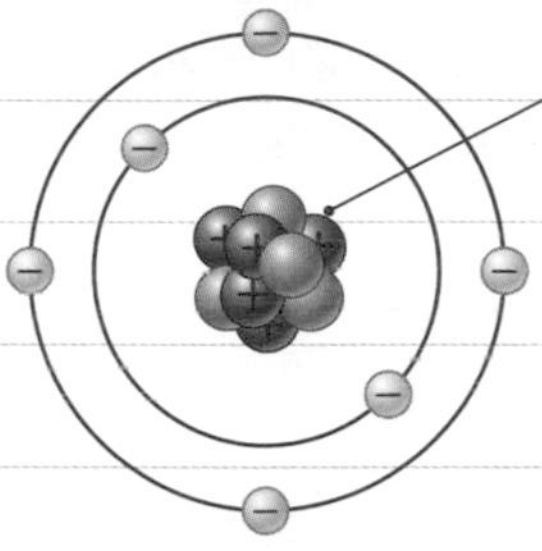

Moving between energy levels

An **electron** will move from a **lower to a higher orbit** if it **absorbs** electromagnetic radiation.

An **electron** will move from a **higher to a lower orbit** if it **emits** electromagnetic radiation.

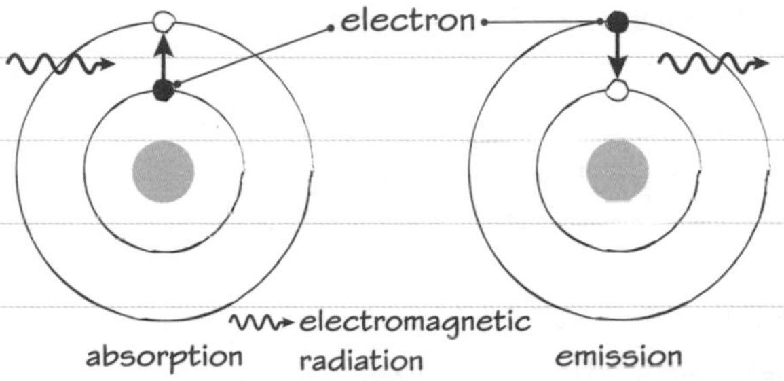

Forming positive ions

Atoms become positively charged particles called **positive ions** when they **lose electrons**. There are now more protons present in the nucleus than there are electrons in the shells.

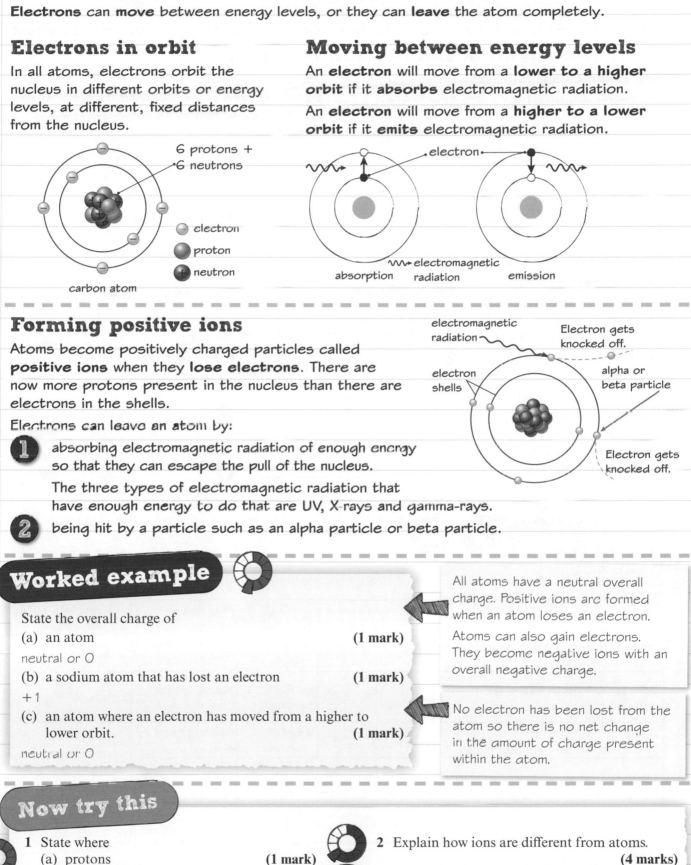

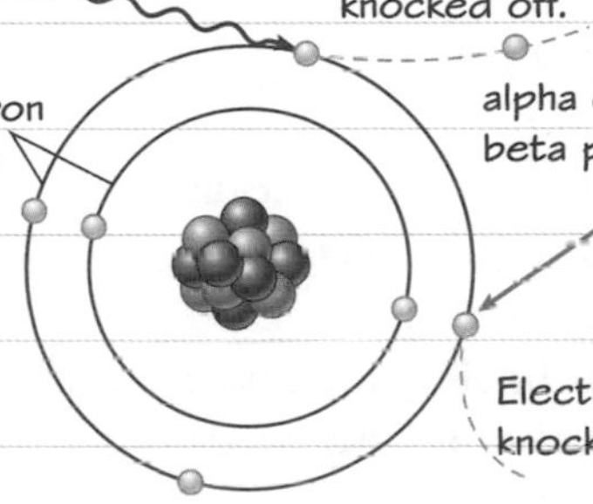

Electrons can leave an atom by:

1. absorbing electromagnetic radiation of enough energy so that they can escape the pull of the nucleus.
   The three types of electromagnetic radiation that have enough energy to do that are UV, X-rays and gamma-rays.
2. being hit by a particle such as an alpha particle or beta particle.

Worked example

State the overall charge of

(a) an atom **(1 mark)**

neutral or 0

(b) a sodium atom that has lost an electron **(1 mark)**

+1

(c) an atom where an electron has moved from a higher to lower orbit. **(1 mark)**

neutral or 0

All atoms have a neutral overall charge. Positive ions are formed when an atom loses an electron.

Atoms can also gain electrons. They become negative ions with an overall negative charge.

No electron has been lost from the atom so there is no net change in the amount of charge present within the atom.

Now try this

1 State where
(a) protons **(1 mark)**
(b) neutrons **(1 mark)**
and
(c) electrons **(1 mark)**
are found in atoms.

2 Explain how ions are different from atoms. **(4 marks)**

3 Suggest how you can tell, by experiment, if an atom has become a positive or a negative ion. **(2 marks)**

Ionising radiation

Alpha, **beta**, **gamma** and **neutron** radiation are **emitted** by **unstable nuclei**. The process is **random**, which means it is not possible to determine exactly **when** any nucleus will decay next.

Alpha, beta, gamma and neutron radiation

Some elements are **radioactive** because their nuclei are **unstable.** This means that they will undergo radioactive decay and change into other elements. Unstable nuclei will decay when alpha, beta, gamma or neutron radiation is emitted.

An **alpha particle** is a helium nucleus. It is composed of two protons and two neutrons. It has a charge of +2 and is the heaviest of the particles emitted by unstable atoms. A **beta particle** is an electron. It has a charge of −1. A **positron** is an anti-electron and has a charge of +1.

A **gamma-ray** is a form of high-energy electromagnetic radiation. It has no mass or charge.

A **neutron** has zero charge.

Properties of radiation

Alpha and beta particles and gamma-rays can collide with atoms, ionising them by causing them to lose electrons.

Neutrons:

- are not directly ionising
- have a very high penetrating power due to them having no charge and not interacting strongly with matter
- can travel through humans and buildings for long distances before being stopped.

1 (α) Alpha particles:
- will travel around 5 cm in air
- very ionising
- can be stopped by a sheet of paper.

2 (β) Beta particles:
- will travel a few metres in air
- moderately ionising
- can be stopped by aluminium
- 3 mm thick.

3 (γ) Gamma-rays:
- will travel a few kilometres in air
- weakly ionising
- need thick lead to stop them.

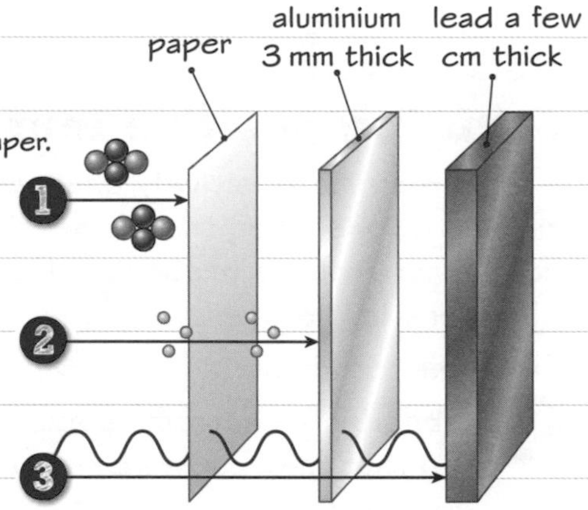

Worked example

Complete the table to show the properties of alpha, beta, gamma and neutron radiation. **(5 marks)**

Type of radiation emitted	Relative charge	Relative mass	Ionising power	Penetrating power	Affected by magnetic fields?
alpha, α (helium nucleus, two protons and two neutrons)	+2	4	heavily ionising	very low, only ~5 cm in air	yes
beta, $\beta-$ (an electron from the nucleus)	−1	1/1840	weakly ionising	low, stopped by thin aluminium	yes, but in the opposite direction to α and $\beta+$
positron, $\beta+$ (a particle with same size as electron but an opposite charge)	+1	1/1840	weakly ionising	low, stopped by thin aluminium	yes
neutron, n	0	1	not directly ionising	high	no, since neither are charged
gamma, γ (waves)	0	0	not directly ionising	very high, stopped only by thick lead	

Now try this

1 Explain why alpha particles are the most ionising radiation. **(3 marks)**

2 Explain why radioactive decay is a random process. **(2 marks)**

Background radiation

Low levels of radiation are present around you all the time. This radiation is both **natural** and **man-made** and is called **background radiation**.

Background radiation

We are always exposed to ionising radiation. This is called **background radiation**. This radiation comes from different sources, as shown in the pie chart.

Radon is a radioactive gas that is produced when **uranium** in rocks decays. Radon decays by emitting an alpha particle. The radon can build up in houses and other buildings. The amount of radon gas varies from place to place, because it depends on the type of rock in the area.

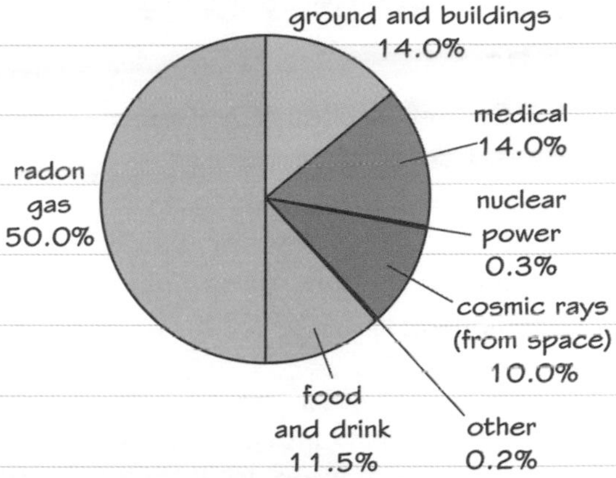

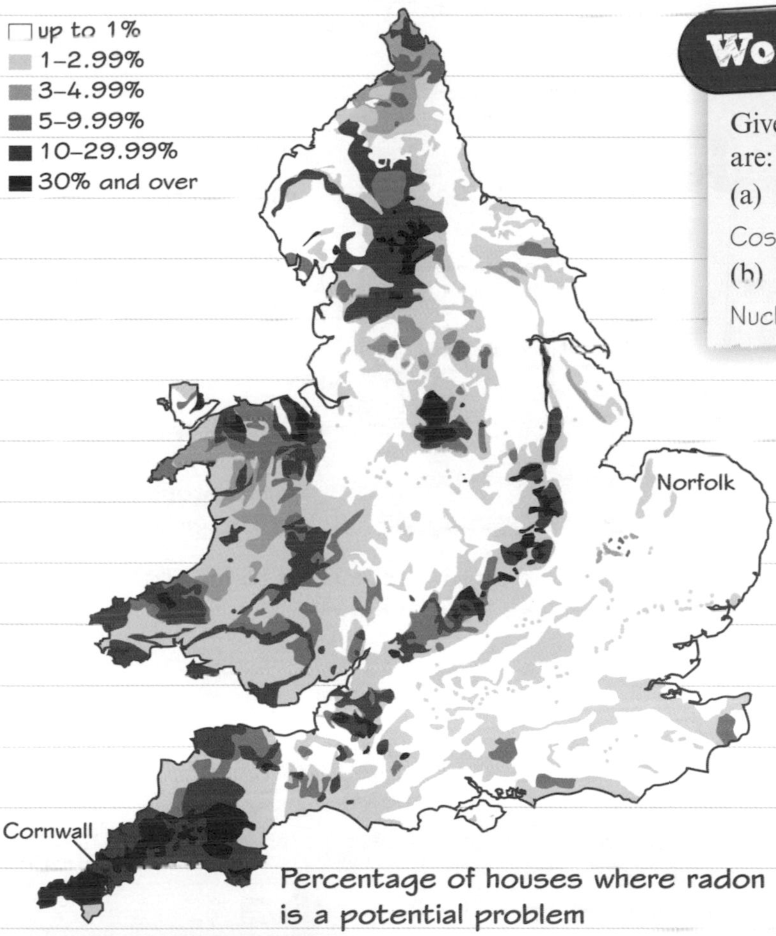

Percentage of houses where radon is a potential problem

Worked example

Give two examples of background radiation that are:

(a) naturally occurring **(2 marks)**

Cosmic rays and radioactivity from the ground.

(b) from human activities. **(2 marks)**

Nuclear power and medical, e.g. hospitals.

Naturally occurring background radiation comes from the environment around you such as the **soil**, **rocks**, **food**, **drink** and **cosmic rays** from outer space.

The artificial sources from human activities come from nuclear power stations, nuclear weapons and from departments in medical settings, such as hospitals, where radioactive materials are made and used.

Now try this

1 Which one of these is a source of artificial background radiation? **(1 mark)**

☐ **A** soil

☐ **B** rock

☐ **C** cosmic rays

☐ **D** medical X-rays

2 Look at the map above. How would the pie chart above change if you lived in
(a) Norfolk
(b) Cornwall? **(2 marks)**

3 Explain why radon is more dangerous inside the body than outside the body. **(3 marks)**

Measuring radioactivity

Photographic film and a **Geiger–Müller tube** can both be used to **measure** and **detect** radiation.

The Geiger–Müller tube

The Geiger–Müller (GM) tube is used to detect nuclear radiation. It is connected to a counter or ratemeter which shows the amount of radiation that has been detected.

The text in red shows what is happening inside the GM tube, but you do not need to remember this in detail.

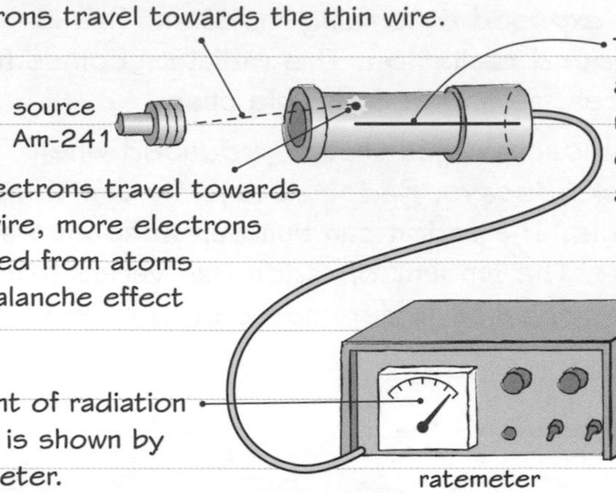

The GM tube contains argon gas. When radiation enters the tube, the atoms of argon are ionised and electrons travel towards the thin wire.
source Am-241
The GM tube has a thin wire in the middle which is connected to a voltage of +400 V.
As the electrons travel towards the thin wire, more electrons are knocked from atoms and an avalanche effect occurs.
The amount of radiation detected is shown by the ratemeter.
ratemeter

Detecting radiation

In 1896, Henri Becquerel discovered that uranium salts would lead to the darkening of a photographic film, even if it was wrapped up so that no light could reach it.

Radiation was being emitted from the uranium nuclei and this was responsible for the darkening of the film. This is now made use of in the nuclear industry as workers will wear a film badge to determine if they are being exposed to different forms of nuclear radiation.

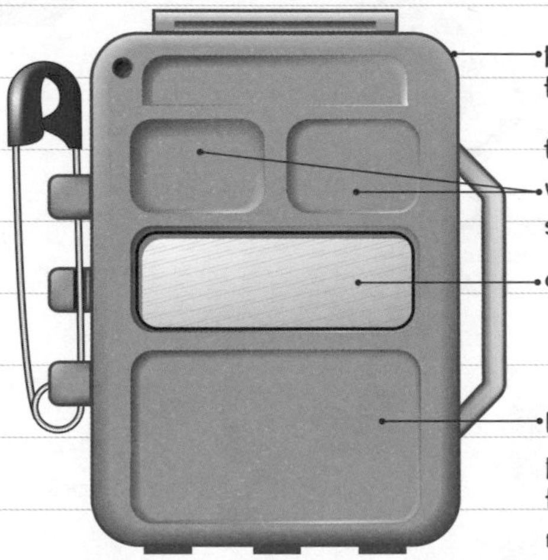

thin and thick plastic windows or aluminium – stop some beta particles

lead between the plastic case and the film – stops beta and most gamma radiation

This dosimeter is a film badge used to monitor the radiation received by its wearer.

Worked example

(a) Explain why there is an open window in the film badge. **(3 marks)**

Light rays will darken the film. Since the film badge is designed to detect nuclear radiation, this acts as a control so that you know that the film is working.

(b) Explain why the film badge has aluminium and lead sheets inserted in it. **(3 marks)**

Aluminium absorbs beta particles and lead absorbs gamma-rays. If there is darkening behind these sheets, then high-energy beta- or gamma-rays may have got through and so the nature of the radiation can be determined and the risk identified.

You should use your knowledge of the penetrating powers of the different types of radioactivity to answer the question.

Now try this

1 Explain why GM tubes are better at detecting alpha particles than beta- or gamma-rays. **(3 marks)**

2 Explain how a film from a film badge would tell you if there was a potential risk to humans. **(3 marks)**

Models of the atom

The **model of the atom** has changed over time, based on **evidence** of its structure that became available from **experiments**.

Plum pudding model

J. J. Thomson's model of the atom was that the atom was like a 'plum pudding' with negatively charged 'electron plums' embedded in a uniform, positively charged 'dough' – a bit like the way currants look when in a Christmas pudding.

This model showed that both positive and negative charges existed in atoms and accounted for the atom being neutral.

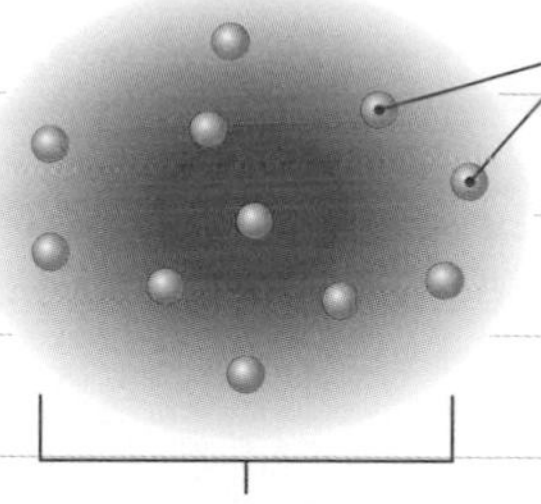

plum pudding model of the atom

Rutherford's model

Rutherford proposed that the atom must contain a very **small**, **positively charged nucleus** which electrons orbit – a bit like planets orbiting the Sun.

Rutherford's hypothesis was proved to be correct by Geiger and Marsden, who fired alpha particles at gold film.

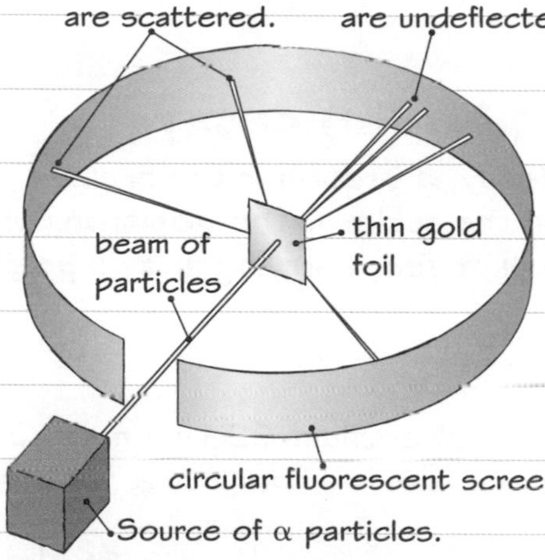

Rutherford scattering

The Bohr model

Niels Bohr showed that **electrons** had to orbit a **positive nucleus** in well-defined **energy levels** or orbits, but could move between energy levels if they gained or lost energy.

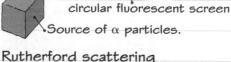

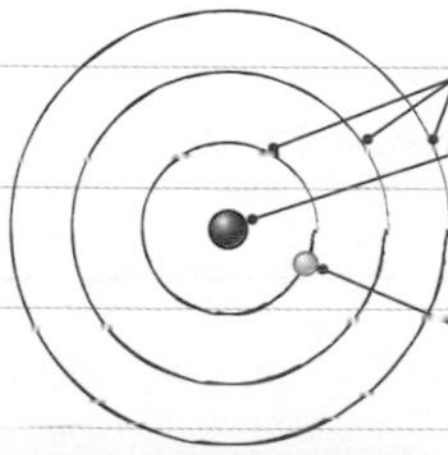

Worked example

Which model of the atom is being described?

(a) The atom is mostly empty space and has a positive nucleus which contains most of the mass of the atom. **(1 mark)**

Rutherford's model

(b) Electrons can only orbit the atom in certain energy levels or orbits. **(1 mark)**

the Bohr model of the atom

(c) The atom is neutral with negative 'plums' embedded in a positive 'dough'. **(1 mark)**

the plum pudding model

Now try this

Refer to page 47 for more information on electron levels.

1 Describe the three models of the atom. **(3 marks)**

2 Explain how Rutherford's work led to the conclusion that the atom had a small, positive nucleus. **(4 marks)**

3 Suggest how the Bohr model allows astronomers to determine which gases are present in the outer surfaces of stars. **(4 marks)**

Beta decay

One way in which unstable nuclei can undergo radioactive decay is by **beta decay.** There are two types – one where an **electron is emitted**, the other where a **positron is emitted**.

Beta-minus (β⁻) decay

In β^- decay, a **neutron** in the nucleus of an unstable atom decays to become a **proton** and an **electron**. The proton stays within the nucleus, but the **electron**, which is the **β^- particle**, is emitted from the nucleus at high speed as a fast-moving electron.

$$n \longrightarrow p + e^-$$

The decay of **carbon-14** into **nitrogen-14** by the emission of a β^- particle is an example of β^- decay. The mass number does not change, but the proton number **increases** by 1.

$$^{14}_{6}C \longrightarrow {}^{14}_{7}N + {}^{0}_{-1}e$$

Beta-plus (β⁺) decay

In β^+ decay, a **proton** in the nucleus decays to become a **neutron** and a **positron**. The neutron stays in the nucleus. The **positron**, which is the **β^+ particle**, is emitted from the nucleus at a very high speed, carrying away a positive charge and a very small amount of the nuclear mass.

$$p \longrightarrow n + e^+$$

The decay of **sodium-22** into **neon-22** by the emission of a positron is an example of β^+ decay. The mass number does not change, but the proton number **decreases** by 1.

$$^{22}_{11}Na \longrightarrow {}^{22}_{10}Ne + {}^{0}_{+1}\beta$$

Worked example

Complete each equation and state whether it is β^- or β^+ decay:

(a) $^{131}_{53}I \longrightarrow {}^{131}_{54}Xe + ?$ **(2 marks)**

Add ${}^{0}_{-1}\beta$. β^- decay

(b) $^{23}_{12}Mg \longrightarrow {}^{23}_{11}Na + ?$ **(2 marks)**

Add ${}^{0}_{1}\beta$. β^+ decay

Uses of beta decay

Carbon-14 is an unstable isotope of carbon and emits β^- particles – it has a half-life of over 5700 years. This means that it can be used for radiocarbon dating, which involves finding the ages of materials that are thousands of years old. Positrons can be used in hospitals to form images of patients by the use of PET scans.

Now try this

1 Describe the changes that take place in the nucleus when it undergoes:

(a) β^- decay **(2 marks)**

(b) β^+ decay. **(2 marks)**

2 How is a β^- particle different from an electron that is found in a stable atom? **(2 marks)**

3 Write a balanced equation for nickel-66 which undergoes β^- decay. **(3 marks)**

Radioactive decay

When unstable nuclei decay, the changes that occur depend on the **radiation** that is emitted from the **nucleus**.

Changes to the nucleus

Type of radiation	Effect on the mass of the nucleus	Effect on the charge of the nucleus
alpha α	nuclear mass reduced by 4 [−4]	positive charge reduced by 2 [−2]
beta $\beta-$	no change [0]	positive charge increased by 1 [+1]
beta $\beta+$	no change [0]	positive charge reduced by 1 [−1]
gamma	no effect on either the mass or the charge of a nucleus	
neutron	mass reduced by 1	no change of nuclear charge

Balancing nuclear decay equations

In any nuclear decay, the total mass and charge of the nucleus are conserved – they are the same before and after the decay. So the masses and charges on each side of the equation must balance.

When uranium-238 undergoes alpha decay, mass and charge are conserved. Nuclei that have undergone radioactive decay also undergo a rearrangement of their protons and neutrons. This involves the loss of energy from the nucleus in the form of gamma radiation. The mass and charge of the gamma-ray must be zero for the equation to balance:

$$^{238}_{92}\text{U} \longrightarrow {}^{234}_{90}\text{Th} + {}^{4}_{2}\alpha + \gamma$$

uranium   thorium   alpha particle   gamma-ray

When neutron decay occurs, mass and charge are also conserved:

$$^{13}_{4}\text{Be} \longrightarrow {}^{12}_{4}\text{Be} + {}^{1}_{0}\text{n}$$

Neutron decay results in another isotope of the same element being formed.

When beryllium decays by neutron emission, the mass and charge are both conserved. The nuclear charge will be +4 on both sides of the decay equation and the total mass will be 13.

Worked example

Balance the nuclear equations for these decays:

(a) Iodine-121 undergoes β^+ decay to form tellurium. **(2 marks)**

$^{121}_{53}\text{I} \longrightarrow {}^{121}_{52}\text{Te} + {}^{0}_{+1}\beta$

$^{0}_{+1}e$ is the emitted β^+ particle.

(b) An isotope of carbon undergoes β^- decay to form nitrogen-14. **(2 marks)**

$^{14}_{6}\text{C} \longrightarrow {}^{14}_{7}\text{N} + {}^{0}_{-1}\beta$

$^{0}_{-1}e$ is the emitted β^- particle.

(c) Radon-220 undergoes α decay to form an isotope of polonium. **(2 marks)**

$^{220}_{86}\text{Rn} \longrightarrow {}^{216}_{84}\text{Po} + {}^{4}_{2}\text{He}$

$^{4}_{2}He$ is the emitted α particle.

Now try this

1 A thorium-232 nucleus undergoes α decay to become radium. Write a balanced equation to show this. **(2 marks)**

2 A nucleus decays by α decay, followed by two β^- decays and then neutron decay. If the original nucleus had a mass number of A and proton number of Z, what would the mass number and the proton number be of the resulting nucleus? **(4 marks)**

Half-life

The **activity** of a radioactive source is the number of atoms that **decay** every second. The unit for activity is the **becquerel (Bq)**. When an atom decays it emits radiation but changes into a more stable isotope.

Unstable atoms

The activity of a source depends on how many **unstable** atoms there are in a sample, and on the particular isotope. As more and more atoms in a sample decay, there are fewer unstable ones left, so the activity decreases. The **half-life** of a radioactive isotope is the time it takes for **half** of the **unstable** atoms to **decay**. This is also the time for the activity to go down by half.

We cannot predict when a particular nucleus will decay. But when there is a very large number of nuclei, the half-life gives a good prediction of the proportion of nuclei that will decay in a given time.

Radioactive decay and half-life

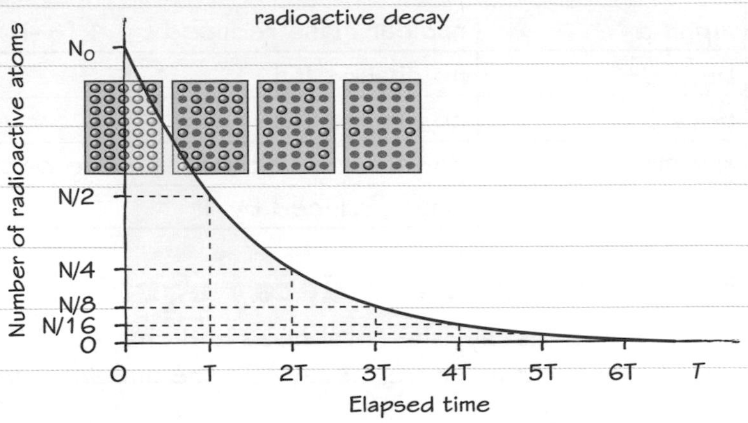

The number of radioactive nuclei present in a sample will halve after each successive half-life. After 1 half-life there will be 50% of the radioactive atoms left; after 2 half-lives there will be 25% of the radioactive atoms left, and so on.

Half-life example

The activity of a radioactive source is 240 Bq and its half-life is 6 hours.

After 1 half-life, the activity will halve to 120 Bq.

After 2 half-lives it will halve again to 60 Bq.

After 3 half-lives it will halve again to 30 Bq.

After 4 half-lives (one day) it will be 15 Bq.

Maths skills

Be careful when multiplying fractions:

After **1 half-life**, $\frac{1}{2}$ of the initial radioactive atoms are left.

After **2 half-lives**, $\frac{1}{2} \times \frac{1}{2}$ or $\frac{1}{4}$ are left.

After **3 half-lives**, $\frac{1}{2} \times \frac{1}{2} \times \frac{1}{2}$ or $\frac{1}{8}$ are left.

Worked example

The graph shows how the activity of a sample changes over 24 hours. What is the half-life of the sample? **(3 marks)**

Activity at time 0 = 1000 Bq

Half of this is 500 Bq. The activity is 500 Bq at 8 hours.

The half-life is 8 hours.

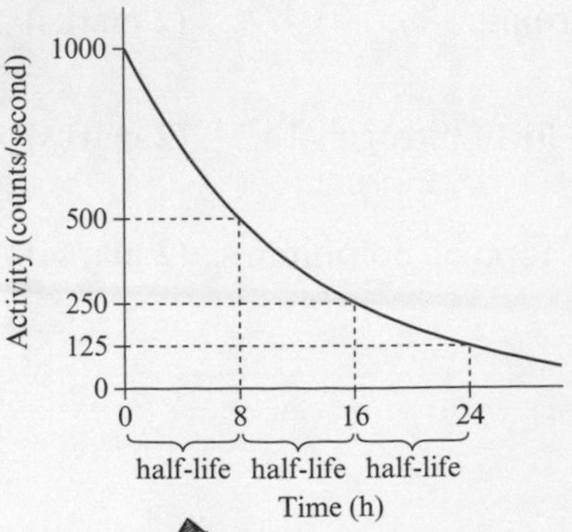

As the activity of a radioactive source decreases, the gradient of the graph will get less and less steep.

Now try this

1 The activity of a source is 120 Bq. Four hours later it is found to be 30 Bq. Calculate the half-life of the source. **(2 marks)**

2 The half-life of a source is 15 minutes. Calculate the fraction that will remain after one hour. **(3 marks)**

Uses of radiation

The **uses** of alpha and beta particles and gamma-rays depend on their **properties**.

Uses of gamma rays

Radiation is used in hospitals to:

- **kill cancer cells** – beams of gamma-rays can be directed at cancer cells to kill them.
- **sterilise surgical instruments** – gamma-rays can be used to sterilise plastic instruments which cannot be sterilised by heating.
- **diagnose cancer** – a **tracer** solution containing a radioactive isotope that emits gamma-rays is injected into the body and taken up by cells which are growing abnormally. The places in the body where the tracer collects are detected with a 'gamma camera'.
- **preserve food** – food irradiated with gamma-rays will last longer as microbes are killed by the high-energy gamma-rays. The food does not become radioactive.

Smoke alarms

Smoke alarms contain a source of **alpha particles**.

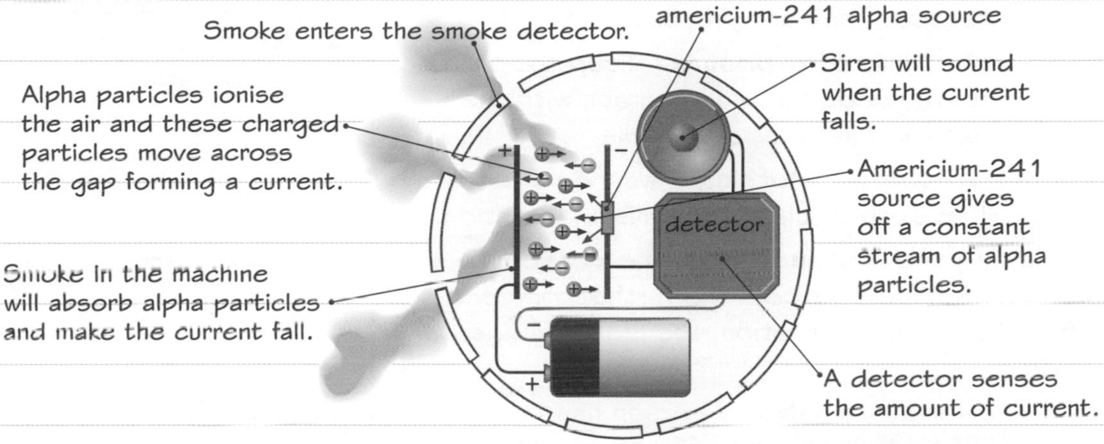

Smoke enters the smoke detector.
americium-241 alpha source
Siren will sound when the current falls.
Alpha particles ionise the air and these charged particles move across the gap forming a current.
Americium-241 source gives off a constant stream of alpha particles.
detector
Smoke in the machine will absorb alpha particles and make the current fall.
A detector senses the amount of current.

Controlling thickness

If the paper is too thick, not as many beta particles get through. → The rollers press together harder to make the paper thinner or move apart to make it thicker.

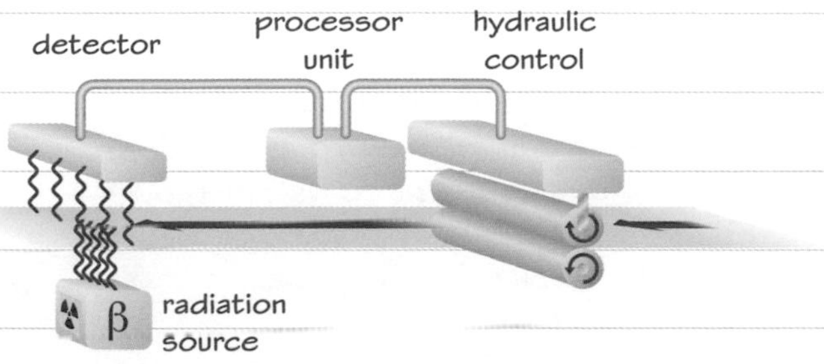

beta particles being used to control the thickness of paper

Worked example

Explain why beta particles are used to control the thickness of paper. **(3 marks)**

Alpha particles would not go through the paper at all. Gamma-rays would pass through the paper too easily, and the amount getting through would hardly change with small changes in the thickness of the paper.

Now try this

1 State three uses of ionising radiation. **(3 marks)**

2 Explain why the radiation in smoke alarms is not dangerous to people in homes. **(3 marks)**

3 Suggest how ionising radiation can be used to find cracks in underground pipes. **(3 marks)**

Dangers of radiation

Ionising radiation can knock electrons out of atoms, turning the atoms into **ions**. This can be very **harmful** to humans.

Ionisation and cellular mutation

The energy transferred by ionising radiation can remove electrons from atoms to form **ions**. Ions are very reactive and can cause mutations to the **DNA** in **cells**. This can lead to **cancer**.

Energy transferred by ionising radiation removes electrons from atoms to form ions. → Ions are reactive and can cause mutations to the DNA in cells. → Damaged DNA can lead to cancer.

Ionising radiation can also cause damage to cell tissue in the form of radiation burns. When its energy is high enough it can also kill cells.

Precautions and safety

People who come into contact with ionising radiation need to be protected. They are protected by:

1. **limiting the time of exposure** – keep the time that a person needs to be in contact with the ionising radiation as **low** as possible.
2. **wearing protective clothing** – wearing a **lead apron** will **absorb** much of the ionising radiation.
3. **increasing the distance from the person to the radioactive source** – the **further** a person is from the ionising radiation, the **less damage** it will do.

To determine how much radiation a person has been exposed to, a film badge may be worn. See page 50 for more details.

The **greater the half-life** of an ionising source, the **longer** it will remain **dangerous**.

Precautions

People using radioactive material take precautions to make sure they stay safe.

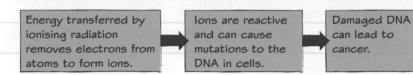

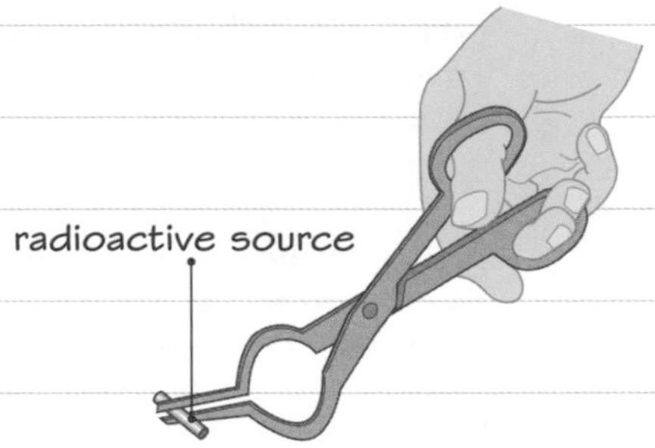

The radioactive source is being moved using tongs to keep it as far away from the person's hand as possible. The source is always kept pointing away from people.

Worked example

Describe two precautions taken by dentists and dental nurses to reduce their exposure to ionising radiation while taking an X-ray. **(2 marks)**

1. They go out of the room in which the X-ray is taking place.
2. They keep the X-ray pulse as short as possible. (This also minimises the patient's exposure.)

The amount of energy that the human body is exposed to from a radioactive source is referred to as the **dose**. The dose needs to be big enough to obtain the X-ray image, but low enough to be safe for the patient and the dentist.

Now try this

1 State two ways in which ionising radiation can be dangerous to humans. **(2 marks)**

2 Explain why ionising radiation is more dangerous than non-ionising radiation. **(3 marks)**

3 Film badges have been superseded in some places by the use of semiconductor devices that measure levels of ionising radiation. Suggest why these devices are preferred to film badges. **(3 marks)**

Contamination and irradiation

You may be exposed to the effects of radioactive materials by being **irradiated** or **contaminated**.

Irradiation

Irradiation is ionising radiation from an external radioactive source travelling to the body – it is not breathed in, eaten or drunk.

Irradiation does not refer to non-harmful rays from televisions, light bulbs or other non-ionising sources.

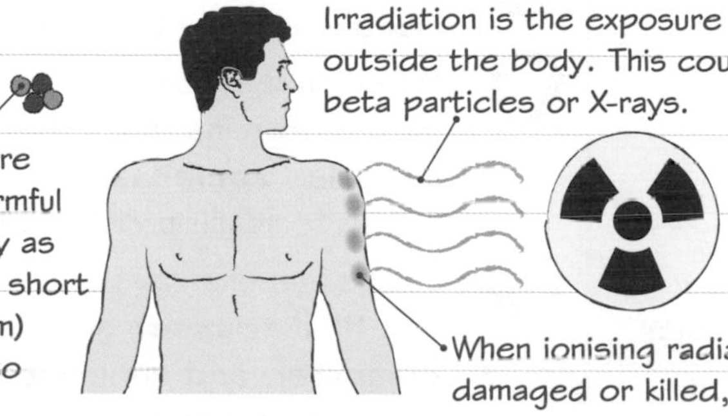

Irradiation is the exposure of a person to ionising radiation from outside the body. This could be in the form of harmful gamma-rays, beta particles or X-rays.
Alpha particles are unlikely to be harmful outside the body as they have a very short range in air (5 cm) and are unlikely to reach a person.
When ionising radiation reaches the body, cells may be damaged or killed, but you will not become radioactive.

Contamination

External contamination occurs when radioactive materials **come into contact with a person's hair, skin or clothing.**
Internal contamination occurs when a **radioactive source is eaten or drunk.** Some nuts, plants, fruits and alcoholic drinks have low levels of radioactivity. This is due to the radioactive minerals that they are exposed to during their growth or manufacture.

Worked example

State what type of contamination the following examples are describing:

(a) eating radioactive strontium-90 that is found in some foods **(1 mark)**

internal contamination

(b) being exposed to cosmic rays from the Sun **(1 mark)**

irradiation

(c) having an X-ray to find if a bone is broken **(1 mark)**

irradiation

(d) radioactive dust comes into contact with the skin **(1 mark)**

external contamination.

Remember that with contamination, the radioactive source comes into contact with the skin or is taken into the body. This can be through the mouth or nose, or through a cut in the skin. With irradiation, the source does not come into contact with the skin.

Now try this

1 Define
(a) irradiation **(2 marks)**
(b) contamination. **(2 marks)**

2 Explain why alpha particles are more likely to cause damage inside the body rather than outside the body. **(3 marks)**

3 Explain why background radiation is a mixture of contamination and irradiation. **(2 marks)**

Medical uses

Ionising radiation can be used **internally** and **externally** to **diagnose illnesses** and **treat cancer**.

Medical tracers

Medical tracers are substances that are used in **biological** processes in the body and contain a **radioisotope**.

The patient can **eat** or **drink** this substance, or it can be **injected** into the body.

The **ionising radiation** emitted by the tracer can be **detected** and the biological process **monitored**. Doctors can diagnose the nature and location of any health problems.

For example, **fluorodeoxyglucose (FDG)** is a radioactive form of glucose that is commonly used as a tracer. Once it is in the blood it travels to the tissues that use glucose. When part of the brain is affected, **less radioactivity** is detected because glucose is not being used.

PET scanners

PET scanners are used to produce 3D colour images of the internal workings of the patient.

1. The **tracer** is a radioactive material that decays quickly by emitting positrons.
2. When these positrons come into contact with electrons in the body, the two particles **annihilate** each other, resulting in the formation of **gamma-rays**.
3. These gamma rays are detected by the **PET scanner** and processed by a computer, and an image is then displayed on a computer screen.
4. As the tracers **decay quickly**, they need to be produced close to where they are used.

Treating tumours internally

Cancer tumours can be treated internally by using a radioactive source that is inside the patient. The source enters the patient by:

1. **injecting** the radioisotope into the patient
2. the patient **eating** or **drinking** something which contains the radioisotope in solid or liquid form.

In the treatment of thyroid cancer, the radioactive element iodine-131 is used. It is swallowed in a capsule. The iodine is taken up by the thyroid gland but not by other parts of the body. This means that the ionising radiation is likely to kill the thyroid cancer without affecting other healthy cells surrounding it.

Treating tumours externally

1. Several beams of gamma rays are fired, from different positions, towards the cancer.
2. Each beam is not energetic enough to kill the tumour, but damages it.
3. By moving the beam, the amount of ionising radiation received by the surrounding tissue is reduced.

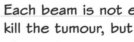

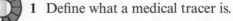

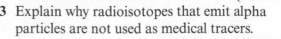

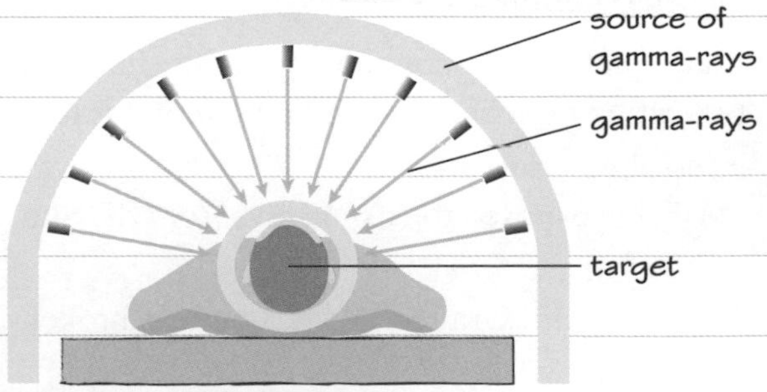

Worked example

Explain why internal treatments are used to treat cancer tumours. **(3 marks)**

When internal treatments are used, the ionising radiation can be targeted at the tumour. This means that most of the ionising radiation goes to the tumour and damage to surrounding areas is reduced. With external treatment, the ionising radiation has to pass through healthy tissue which can be damaged.

Now try this

1 Define what a medical tracer is. **(2 marks)**

2 Explain why several low-energy gamma-ray beams are used instead of a single high-energy beam when treating cancer with gamma-rays from an external source. **(4 marks)**

3 Explain why radioisotopes that emit alpha particles are not used as medical tracers. **(3 marks)**

Nuclear power

Electrical energy can be generated in power stations that use **nuclear fuel**. There are advantages and disadvantages of nuclear power.

Energy from the nucleus

All nuclear reactions can be a source of energy:

1. **Radioactive decay** can be in the form of alpha or beta particles, or gamma rays (which are electromagnetic waves).
2. **Nuclear fusion:** the fusing of hydrogen nuclei in the Sun releases enormous amounts of light and heat, some of which we receive on Earth.
3. **Nuclear fission:** uranium nuclei are split by slow-moving neutrons in a nuclear reactor, resulting in the release of enormous amounts of thermal energy.

Nuclear power and waste

Nuclear power is a **highly efficient** way of producing electrical energy. Much less nuclear fuel is needed, compared with coal, oil or gas, to produce the same amount of electrical energy. However, nuclear power stations do produce radioactive **nuclear waste**, which needs to be dealt with. Some of it is radioactive for thousands of years.

Nuclear power

There are advantages and disadvantages to using nuclear energy to generate electricity.

- 👍 Nuclear power stations do not produce **carbon dioxide**, so they do not contribute to climate change.
- 👍 Supplies of nuclear fuel will **last longer** than supplies of fossil fuels.

However, construction processes produce carbon dioxide, so carbon dioxide will be added to the atmosphere when the power station is built and when fuel rods are made.

- 👎 It is difficult and expensive to **store** nuclear waste safely.
- 👎 An **accident** in a nuclear power station can spread **radioactive material** over a large area.
- 👎 Many people think that nuclear power is dangerous, and do not want new nuclear power stations to be built.

Nuclear power stations do not make the local area more radioactive when they are working properly. This only happens if there is an accident.

Worked example

Explain the disadvantages of nuclear power. **(3 marks)**

Nuclear radiation is very dangerous, so a nuclear accident could kill many people. There is much radioactive waste which is difficult and expensive to store, and some of it is radioactive for many thousands of years. Public perception surrounding nuclear power as a form of energy means that people often distrust it because of previous accidents such as at Chernobyl and Fukushima.

Despite the problems, it is important to remember that nuclear power has some real advantages, including the fact that no carbon dioxide is produced when a power station is in operation (and so they do not affect climate change) and that the fuel supplies will last longer than coal, oil or gas reserves.

Now try this

1 'Nuclear power does not produce carbon dioxide emissions.' Explain why this statement is not correct. **(3 marks)**

2 Describe how nuclear power pollutes the environment. **(4 marks)**

Nuclear fission

The fission of uranium-235 results in the **release of a large quantity of energy**, which is then used to **heat water** in power stations.

Nuclear fission

In a **fission reaction**, a large unstable nucleus splits into two smaller ones. For example, a **uranium-235** nucleus splits up when it absorbs a **neutron**. The fission of uranium-235 produces two **daughter nuclei**, two or more neutrons, and also releases energy. The daughter nuclei are also radioactive.

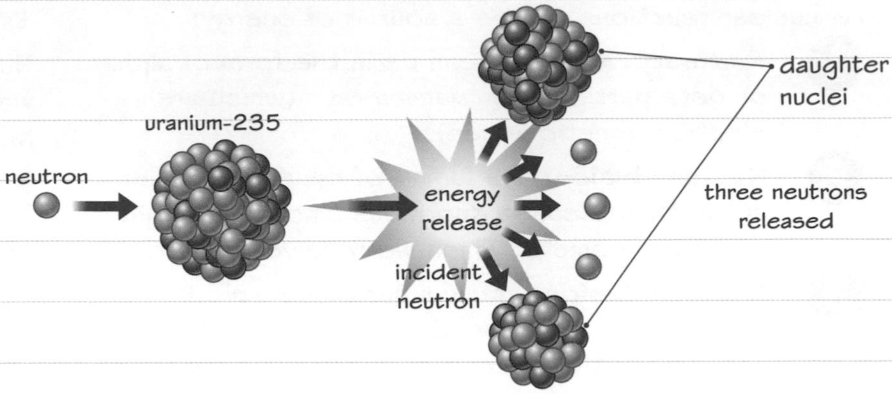

Controlled chain reactions

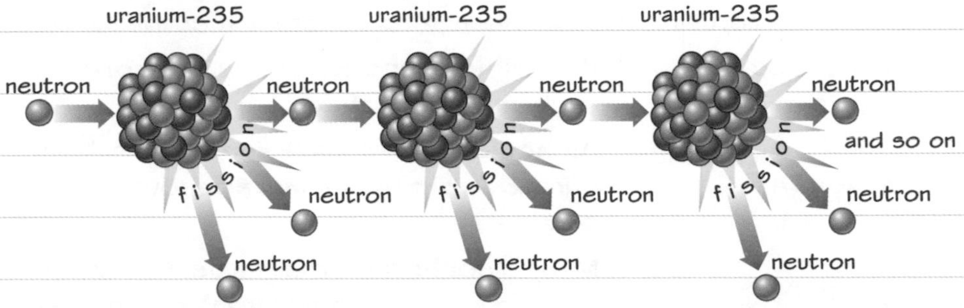

Two of the neutrons are absorbed by other materials. Only one neutron from each fission can cause other fission. This is a **controlled chain reaction**.

Chain reactions

The neutrons released by the fission of U-235 may be absorbed by other nuclei. Each of these nuclei may undergo fission, and produce even more neutrons. This is called a **chain reaction.** If a chain reaction is not controlled there will be a nuclear explosion.

Nuclear reactors make use of **controlled chain reactions** (see next page).

Worked example

Describe how a chain reaction can be controlled. **(3 marks)**

A chain reaction can be controlled by using a different material to absorb some of the neutrons. This slows the reaction down because there are fewer neutrons to cause more nuclei to undergo fission.

Now try this

1 Describe what happens in the fission of uranium-235. **(3 marks)**

2 Explain the difference between a chain reaction and a controlled chain reaction. **(4 marks)**

3 Explain why:

(a) neutrons in a nuclear reaction need to be slow moving **(2 marks)**

(b) only one neutron should be absorbed by a uranium-235 nucleus. **(2 marks)**

Nuclear power stations

You need to know how electricity can be generated from **nuclear fission** in nuclear power stations.

Generating electrical energy from nuclear fission

It is the thermal energy from nuclear fission that is most useful for the generation of electricity in a power station.

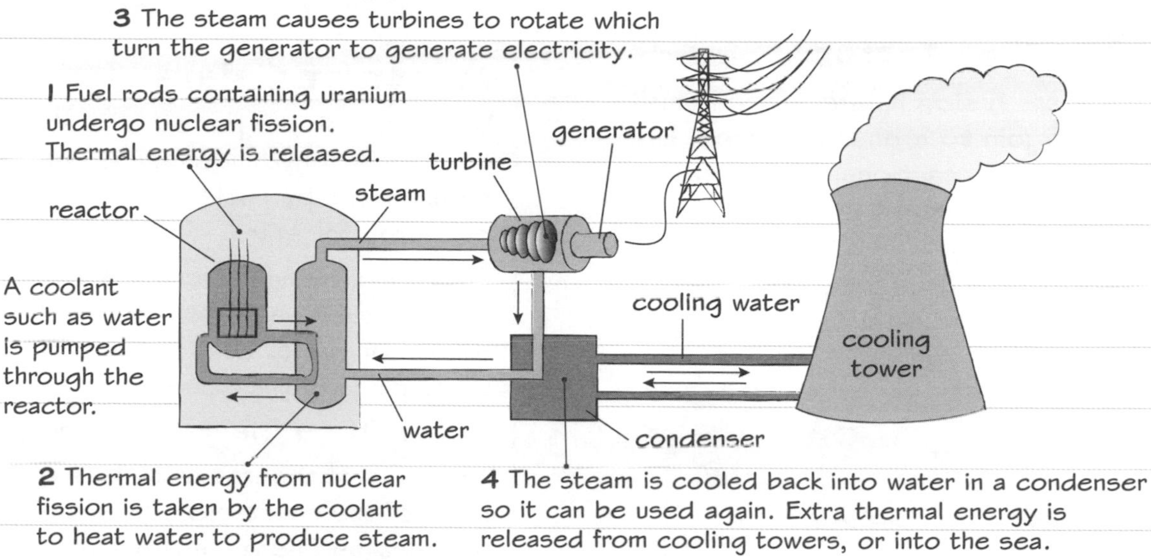

3 The steam causes turbines to rotate which turn the generator to generate electricity.
1 Fuel rods containing uranium undergo nuclear fission. Thermal energy is released.
generator
turbine
steam
reactor
A coolant such as water is pumped through the reactor.
cooling water
cooling tower
water
condenser
2 Thermal energy from nuclear fission is taken by the coolant to heat water to produce steam.
4 The steam is cooled back into water in a condenser so it can be used again. Extra thermal energy is released from cooling towers, or into the sea.

Controlling the chain reaction

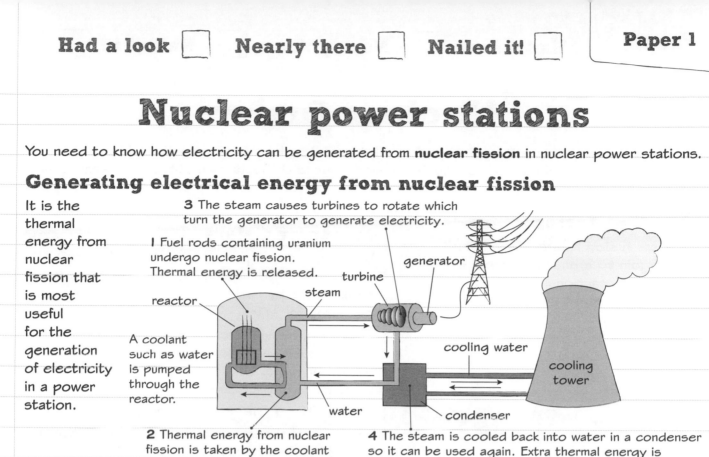

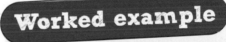

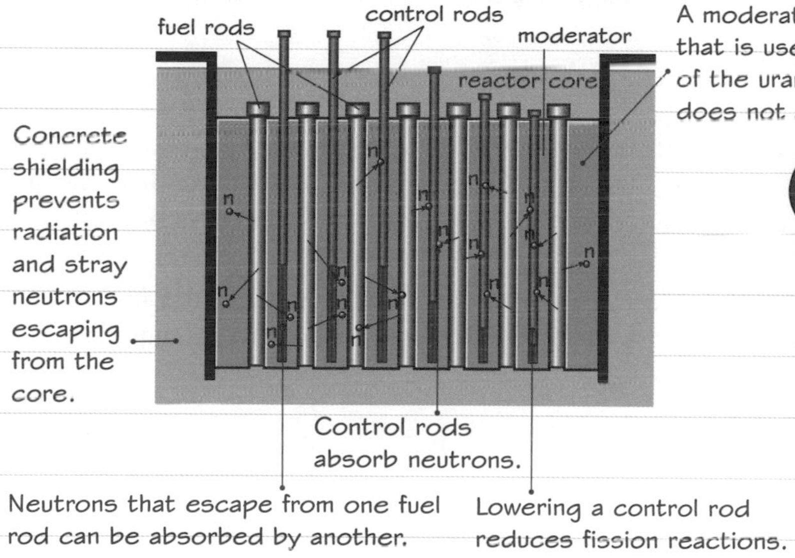

Concrete shielding prevents radiation and stray neutrons escaping from the core.

Neutrons that escape from one fuel rod can be absorbed by another.

A moderator is a material (often graphite or water) that is used to slow down the neutrons so that fission of the uranium-235 nuclei can occur. The moderator does not absorb neutrons.

Worked example

Explain how the control rods work in a nuclear power station. **(3 marks)**

The control rods **absorb** neutrons. If the control rods are pushed down into the core, more neutrons are absorbed and the chain reaction slows down. If they are pulled out, fewer neutrons are absorbed and the chain reaction speeds up.

Worked example

(a) Describe how nuclear fuel is used in a power station. **(2 marks)**

Nuclear fuel undergoes fission to release huge quantities of thermal energy. This is used to heat water to produce steam.

(b) Describe what happens to the products of nuclear fission. **(2 marks)**

The radioactive daughter products need to be disposed of safely. They are usually stored until the activity has decreased to a safe level.

Now try this

1 Explain how pulling control rods out of the core of a nuclear reactor can increase the amount of thermal energy released by the reactor. **(4 marks)**

2 Evaluate the use of nuclear fuel to produce electrical energy compared with fossil fuels. **(6 marks)**

Nuclear fusion

Nuclear fusion involves the **joining of smaller nuclei** to form **larger nuclei**. Obtaining a continuous supply of energy from nuclear fusion is **more difficult** than obtaining energy from nuclear fission.

Nuclear fusion

Nuclear **fusion** happens when small nuclei join to form larger ones. Like all nuclear reactions, fusion reactions **release energy**.

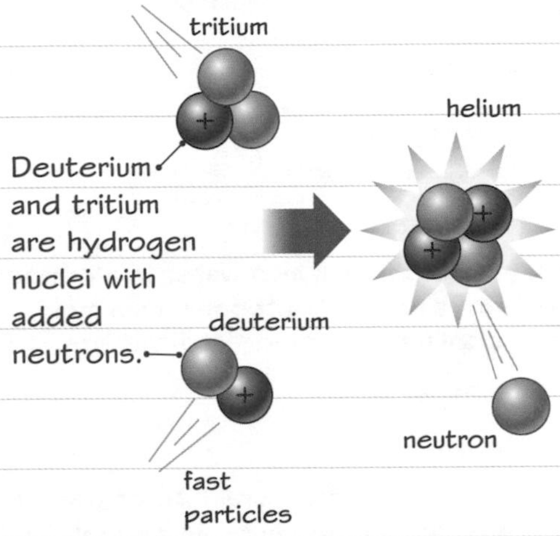

Fusion and stars

During the stable period of a star's life, vast quantities of hydrogen nuclei are converted to helium nuclei by nuclear fusion. This is the source of energy for stars.

Over time, heavier elements are formed and the star will eventually die. This will take our Sun billions of years. One of the typical reactions that takes place in the Sun is:

$$^{2}_{1}H + ^{3}_{1}H \longrightarrow ^{4}_{2}He + ^{1}_{0}n + \text{energy}$$

The mass of the products is slightly less than the mass of the reactants – this mass difference is released as energy in the form of thermal energy.

This reaction takes place under very high pressure and temperature in stars.

Difficulties of fusion

Nuclei need to get very close to each other before fusion can happen. Under normal conditions the positive charges on nuclei repel each other. This is called **electrostatic repulsion**. Only at very high temperatures and pressures are the nuclei moving fast enough for them to overcome this electrostatic repulsion.

The very high temperatures and pressures needed are very difficult to produce in a fusion power station. All the experimental fusion reactors built so far have used more energy than they have produced!

Worked example

Explain the difference between nuclear fusion and nuclear fission. **(3 marks)**

In nuclear fusion hydrogen nuclei fuse to produce helium nuclei whereas in nuclear fission uranium-235 splits into two smaller nuclei and two or three neutrons.

Nuclear fusion requires very high temperatures and pressures whereas nuclear fission requires a slow-moving neutron to be absorbed.

Both processes release large quantities of energy.

Fusion can occur in the Sun and other stars because the enormous temperatures and pressures allow the charged nuclei to overcome the repulsive electrostatic forces between them.

Now try this

1 Describe what happens during nuclear fusion. **(3 marks)**

2 Explain why there are no commercial power stations using fusion reactions. **(3 marks)**

3 Explain how the processes of fusion and fission release energy. **(4 marks)**

Extended response – Radioactivity

There will be one or more 6 mark questions on your exam paper. For these questions, you will need to think scientifically and structure your answer logically, showing how the points you make are related to each other. You can revise the topics for this question, which is about **radioactivity**, on pages 45–62.

Worked example

Some types of cancer can be treated by using a radioactive material from inside the body, whereas other cancers are treated by using a radioactive source outside the body.

Explain how cancer can be treated in both cases.

Your answer should refer to examples of suitable isotopes, the radiation emitted and the half-life of these isotopes. **(6 marks)**

Internal radiation therapy involves the placing of seeds, ribbon or capsules inside the patient's body near to the tumour. This is called brachytherapy. Alternatively, the radioisotope can be injected, swallowed or enter the body via an intravenous drip. The radioisotope then travels through the body, locating and killing cancer cells. For radioactive materials that are used internally to treat cancer, the half-life of the source is low – it could be a few hours, days or weeks depending on the nature of the cancer being treated. One example is iodine-131, which has a half-life of 8 days and is used to treat thyroid cancer. It is a beta emitter, although alpha emitters are also used internally as these only need to travel short distances. The half-life of the source needs to be long enough to treat the cancer but not so long as to cause long-term harm to the patient.

When treating a tumour externally, weak gamma-rays are fired from different positions around the body so as to focus the gamma-rays on the cancer tumour but not to cause too much damage to the healthy cells surrounding the tumour. The gamma-rays come from an external source which should have a long half-life so that it does not need to be constantly replaced. An example of such a source is cobalt-60, which has a half-life of between 5 and 6 years.

Command word: Explain

When you are asked to **explain** something, it is not enough just to state or describe it. Your answer **must** contain some reasoning or justification of the points you make. Your explanation **can** include mathematical explanations, if calculations are needed.

The answer starts by explaining how a cancer tumour can be targeted from inside the body. The methods used to get the material into the body are explained, as is an example of a suitable isotope, a typical half-life value and suitable types of radioactive particle that are emitted.

The second part of the answer refers to how cancer can be treated from an external source. The reason for the long half-life and a suitable gamma source are given.

Internal vs external sources of radiotherapy

External sources of radiotherapy tend to use gamma-rays for irradiating the patient. The patient does not become radioactive after treatment. However, a patient does become radioactive if they are treated internally as they have been contaminated with radioactive material which can emit particles that will leave the body.

Now try this

Radioactive material is used in smoke alarms. Explain the characteristics of the radioactive material used in smoke alarms. Your answer should refer to suitable and unsuitable radioactive materials for this purpose. **(6 marks)**

The Solar System

The **Solar System** contains the **Sun**, **eight planets**, their **natural satellites**, **dwarf planets**, **asteroids** and **comets**. Ideas about the structure of the Solar System have changed over time.

The Solar System

You need to remember the order of the eight planets, going from the Sun.

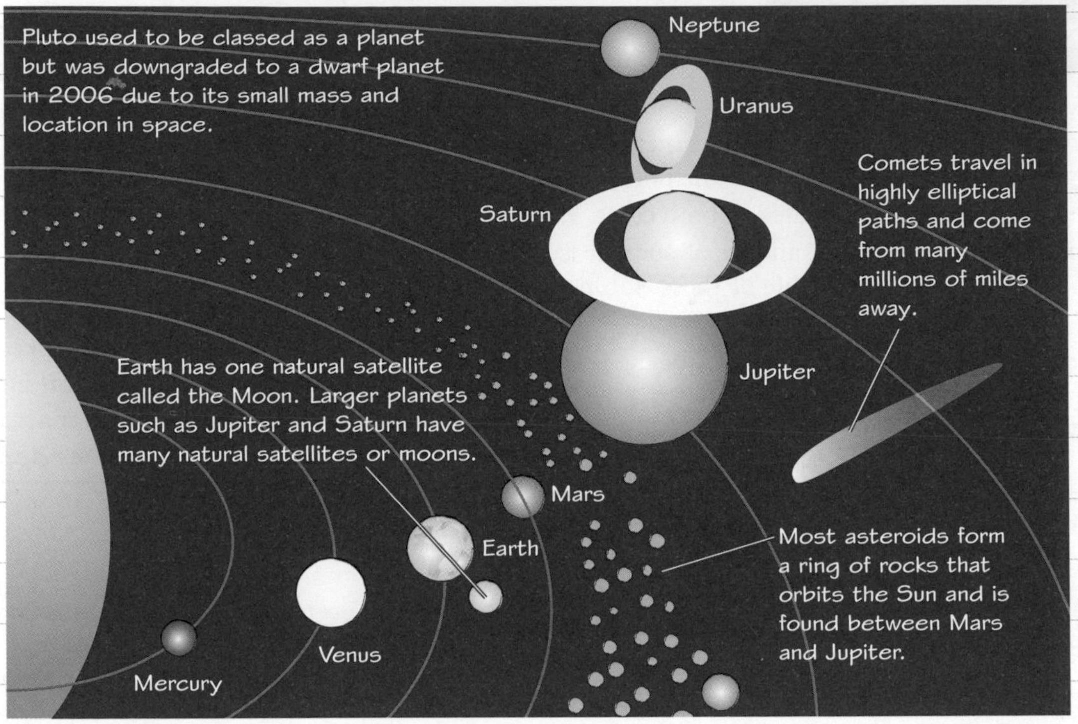

Pluto used to be classed as a planet but was downgraded to a dwarf planet in 2006 due to its small mass and location in space.
Neptune
Uranus
Saturn
Comets travel in highly elliptical paths and come from many millions of miles away.
Jupiter
Earth has one natural satellite called the Moon. Larger planets such as Jupiter and Saturn have many natural satellites or moons.
Mars
Earth
Most asteroids form a ring of rocks that orbits the Sun and is found between Mars and Jupiter.
Venus
Mercury

Changing ideas

Many early people thought that the **Sun** and all the **planets** moved around the Earth.

In about 130 CE a Greek astronomer called Ptolemy published a **geocentric** ('Earth-centred') model that tried to explain the motions of the planets.

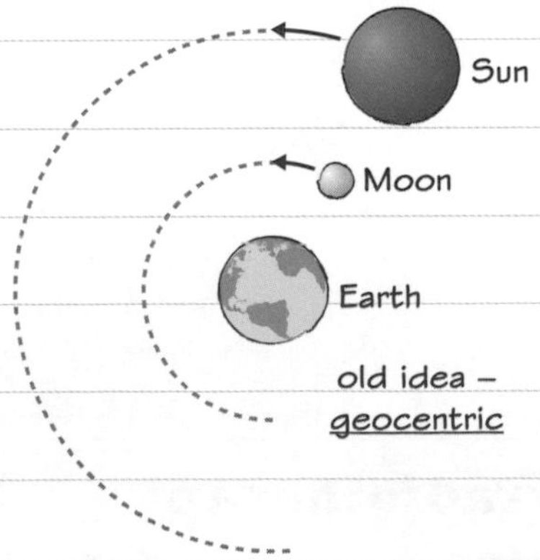

In 1543 Nicolaus Copernicus published a book that argued that the Earth and the planets were in **orbits** around the Sun. This is a **heliocentric** ('Sun-centred') model.

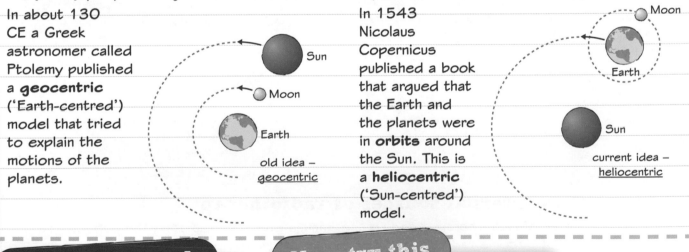

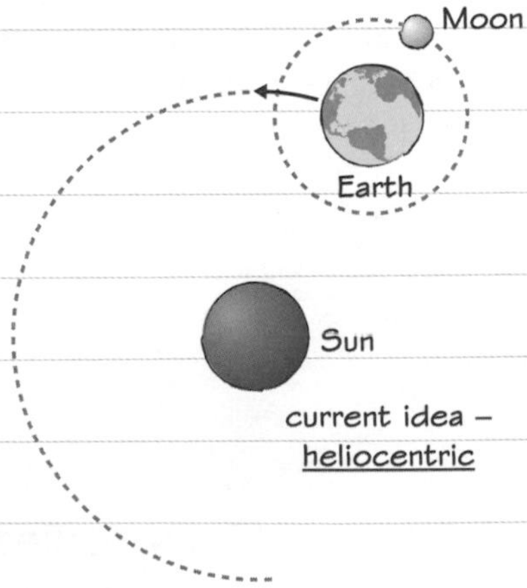

Worked example

State what the Solar System is composed of. **(3 marks)**

The Solar System is composed of the Sun, eight planets, asteroids, comets and dwarf planets such as Pluto.

Now try this

1 State the difference between a geocentric and a heliocentric model of the Solar System. **(2 marks)**

2 Name the planets in order from the Sun, starting with the furthest. **(2 marks)**

3 Suggest why some planets have been discovered more recently than others. **(3 marks)**

You could make up a mnemonic to help you remember the order.

Satellites and orbits

Bodies of **lower mass** will travel in **orbits** around **bodies of much higher mass.** The nature of these orbits needs to be known for a variety of natural and artificial satellites.

Natural satellites

Natural satellites were formed by natural processes. Examples include:

- the eight **planets** in the Solar System which orbit the Sun
- the **moons** which orbit planets in the Solar System
- **comets** which orbit the Sun.

Artificial satellites

Artificial satellites are manufactured and have been launched into space from Earth by using rockets. Examples include:

- satellites in **geostationary orbits** around the Earth being used for global positioning satellite systems (GPS)
- satellites in **low polar orbits** around the Earth being used for weather monitoring, military, spying or Earth observation purposes
- satellites **sent from Earth** to orbit and monitor the Sun, other planets or asteroids in the Solar System.

Type of orbit

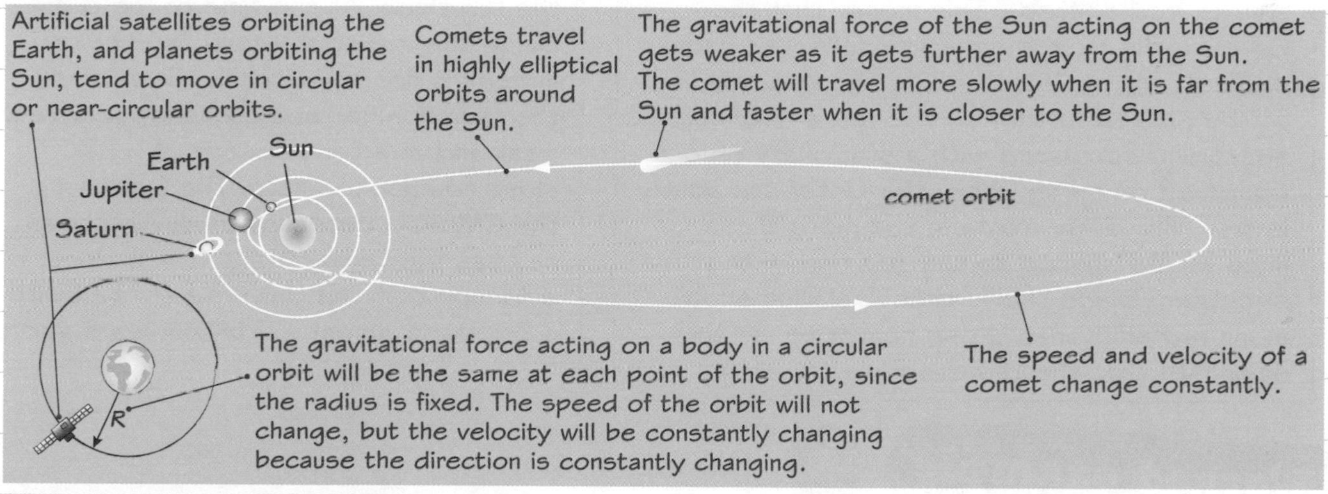

Artificial satellites orbiting the Earth, and planets orbiting the Sun, tend to move in circular or near-circular orbits.
Comets travel in highly elliptical orbits around the Sun.
The gravitational force of the Sun acting on the comet gets weaker as it gets further away from the Sun. The comet will travel more slowly when it is far from the Sun and faster when it is closer to the Sun.
Earth
Sun
Jupiter
Saturn
comet orbit
R
The gravitational force acting on a body in a circular orbit will be the same at each point of the orbit, since the radius is fixed. The speed of the orbit will not change, but the velocity will be constantly changing because the direction is constantly changing.
The speed and velocity of a comet change constantly.

Worked example

Compare the motion of Earth, Jupiter and Saturn as they orbit the Sun. **(3 marks)**

All three planets travel in circular orbits and have a constant speed and a changing velocity. As the distance from the Sun increases, the orbital speed decreases. The Earth has a higher orbital speed than Jupiter which has a higher orbital speed than Saturn.

The planets in the Solar System are in stable orbits. For example:

1. If the Earth's orbital speed increased then the gravitational force from the Sun would not keep it in orbit. It would fly out of its orbit, into a new orbit of greater radius.
2. If the Earth's orbital speed decreased, the Sun's gravitational field would cause it to fall towards it, into a new orbit of smaller radius.

Now try this

1 State the nature of the speed and velocity of a body that is moving in a circular orbit. **(2 marks)**

2 Explain how the speed and velocity of a comet changes over the course of one orbit. **(4 marks)**

3 Explain what might happen to a GPS satellite if its orbital speed changed. **(6 marks)**

Theories about the Universe

Scientists' ideas about the **birth** and **evolution** of the **Universe** have changed over time.

Existence theories

There are two different **theories** about the Universe.

The **Big Bang** theory says that the whole Universe started out as a tiny particle about **13.8 billion years** ago. The Universe expanded from this point in space. The Universe is still **expanding** today.

The **Steady State** theory says that the Universe has always existed. It is expanding, and new matter is being created as it expands.

The Big Bang theory is currently the most accepted model, based on the available evidence.

Red-shift

When objects that emit light, such as galaxies, are moving away from us then the light they emit is 'red-shifted'. This means that the wavelengths from them when detected back on Earth will be 'stretched'. The spectral lines will be shifted further to the red end of the spectrum, compared with a stationary object observed on Earth. Also, the faster the galaxy is travelling away from us, the more these lines will be shifted to the red end of the spectrum. Since the furthest galaxies show more red shift than those nearer to us, this suggests that the Universe is expanding.

CMB

The CMB (**Cosmic Microwave Background** radiation) is the 'echo' of the Big Bang. At the time of the Big Bang, this radiation would have been incredibly hot and intense. Over time, the radiation has cooled and is now very weak and at a temperature of −270°C, close to absolute zero. The fact that the CMBR is detected in all directions and has the temperature that it does, is evidence that the Universe started with a Big Bang almost 14 billion years ago.

Worked example

Describe the cosmic microwave background radiation (CMB) and explain why the Big Bang theory is the currently accepted model for the beginning of the Universe. **(4 marks)**

Although red-shift is evidence for both the Steady State theory and the Big Bang theory, it is the cosmic microwave background radiation that provides the key evidence for the Big Bang theory.

The cosmic microwave background radiation is detected by radio telescopes and comes from all over the sky. The Big Bang theory says that this radiation was released at the beginning of the Universe.

Red-shift supports both the Big Bang and the Steady State theories, but the CMB only supports the Big Bang theory. The Big Bang theory is accepted because it has the most evidence supporting it.

Now try this

1 What evidence is there to support the Big Bang and the Steady State theories? **(1 mark)**

2 Why is the cosmic microwave background radiation such important evidence? **(3 marks)**

3 Distances in space are sometimes measured in light years. Galaxy NGC 55 is 7.08 million light years away. The Pinwheel galaxy is 21 million light years away.
Explain the similarities and differences you would expect in the light from these galaxies as a result of their movement. **(4 marks)**

Doppler effect and red-shift

The **Doppler effect** and **red-shift** both provide **evidence** for the nature of the Universe.

The Doppler effect

When a wave source is moving relative to an observer, there is a change in the observed frequency and wavelength of the waves. Consider the siren in the car shown in the picture.

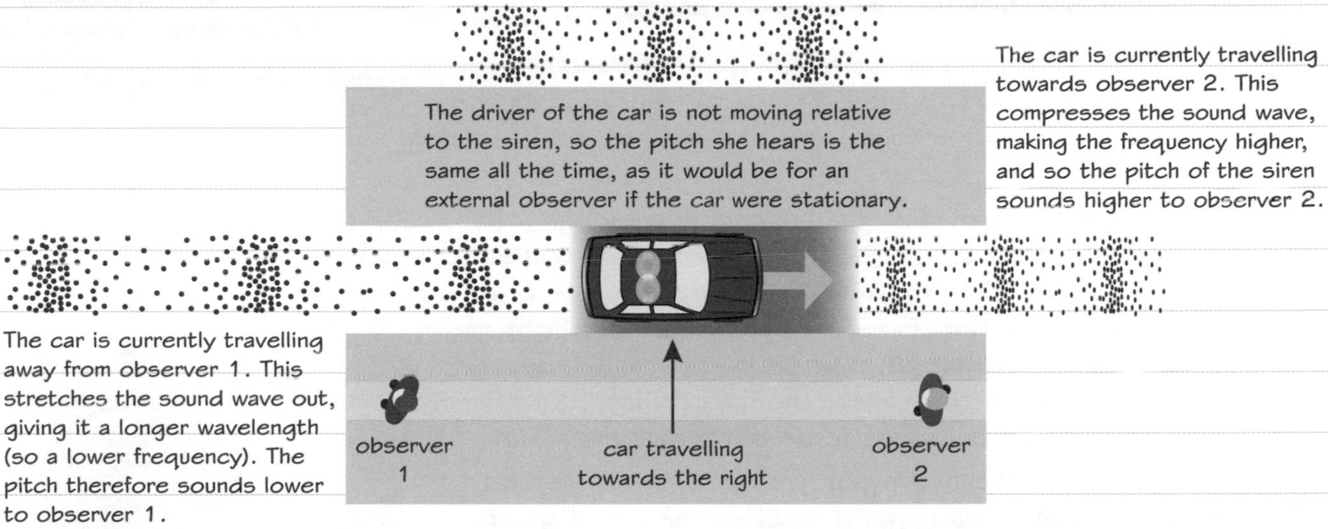
The driver of the car is not moving relative to the siren, so the pitch she hears is the same all the time, as it would be for an external observer if the car were stationary.
The car is currently travelling towards observer 2. This compresses the sound wave, making the frequency higher, and so the pitch of the siren sounds higher to observer 2.
The car is currently travelling away from observer 1. This stretches the sound wave out, giving it a longer wavelength (so a lower frequency). The pitch therefore sounds lower to observer 1.
observer 1
car travelling towards the right
observer 2

Red-shift and the Doppler effect

The Doppler effect occurs for light waves as well as for sound waves.

A light source moving away from an observer will have a **greater wavelength** and a **lower frequency**. It is **'red-shifted'**.

A light source moving towards an observer will have a **smaller wavelength** and a **higher frequency**. It is **'blue-shifted'**.

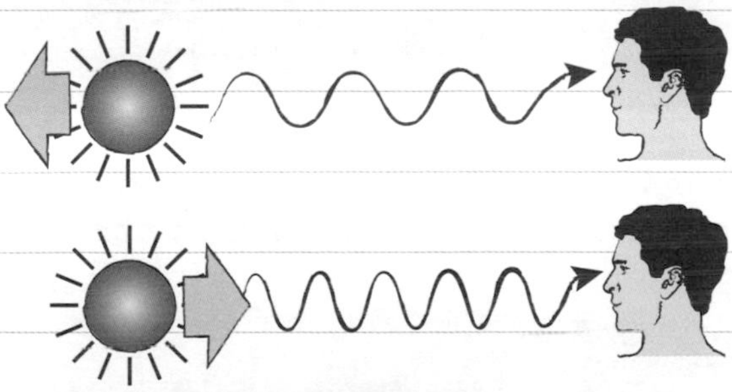

Red-shift and evidence for the expanding Universe

In 1929, the astronomer Edwin Hubble discovered that the light reaching Earth from galaxies was red-shifted. He found this out by looking at the spectral lines in the absorption spectra of stars that are moving away from us.

Galaxy b's lines are red-shifted the most, so it is moving away from us the fastest.

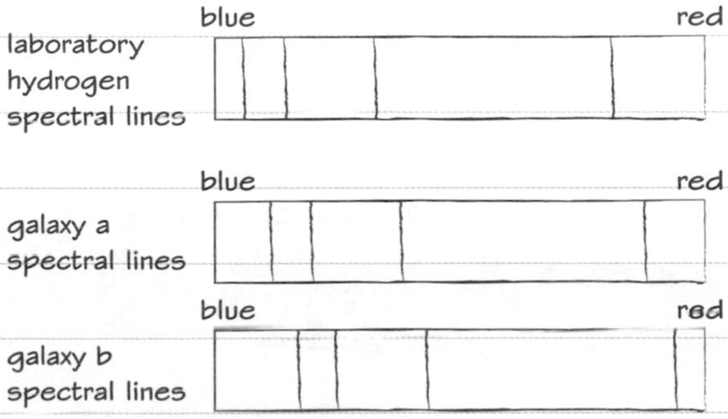
blue
red
laboratory hydrogen spectral lines
blue
red
galaxy a spectral lines
blue
red
galaxy b spectral lines

Worked example

Explain how red-shift suggests that the Universe is expanding. **(4 marks)**

The light received on Earth from most galaxies is red-shifted. This means that most galaxies are moving away from us. The furthest galaxies from us have the greatest red-shift, so are moving away from us faster. This suggests that the Universe is expanding.

Now try this

1 Describe an example of the Doppler effect for
(a) sound waves **(2 marks)**
(b) light waves. **(2 marks)**

2 Suggest how the speed of a galaxy can be determined from the absorption lines in its spectrum. **(3 marks)**

Life cycle of stars

Stars go through a number of **stages** in their lives. Their eventual **fate** depends on their **mass**.

Low-mass stars

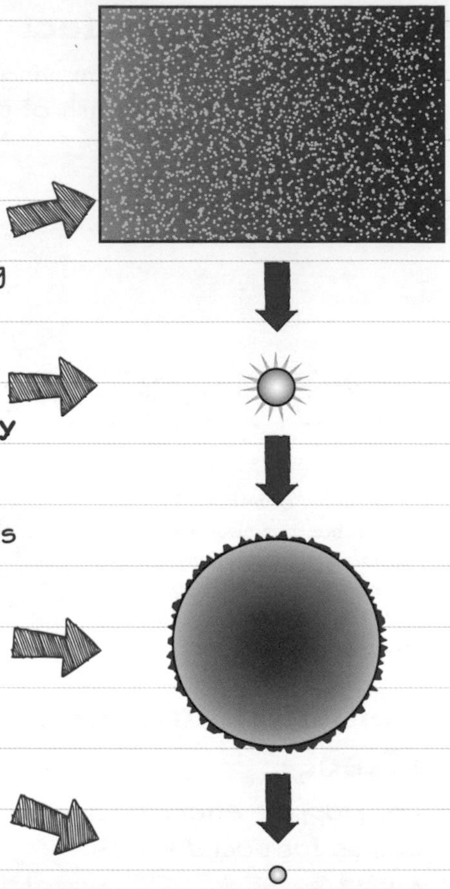

The Sun was formed about 4.6 billion years ago and is nearly halfway through its life. Stars with masses up to four times the mass of the Sun are classified as low-mass stars.

1. The cloud of **dust** and **hydrogen gas** (a **nebula**), is pulled inwards by the force of gravity. As the dust and gas contracts the nebular gets hotter since work is being done on it.
2. Eventually, the dust and gas become hot enough for the hydrogen nuclei to fuse. **Nuclear fusion** leads to heavier **helium nuclei** being produced and large amounts of **energy** being released. The star will begin to give out **light** and is now a **main-sequence** star. Stars are main-sequence stars for most of their lives. The inwards force of gravity is balanced by the outwards force of thermal expansion.
3. When most of the hydrogen gas has been converted to helium, the star will **expand** and become a **red giant**. When the core of this star collapses, other **heavier elements** are formed.
4. Eventually, all nuclear fusion stops due to the elements that cause fusion being used up, and the star **collapses** to become a **white dwarf**.

High-mass stars

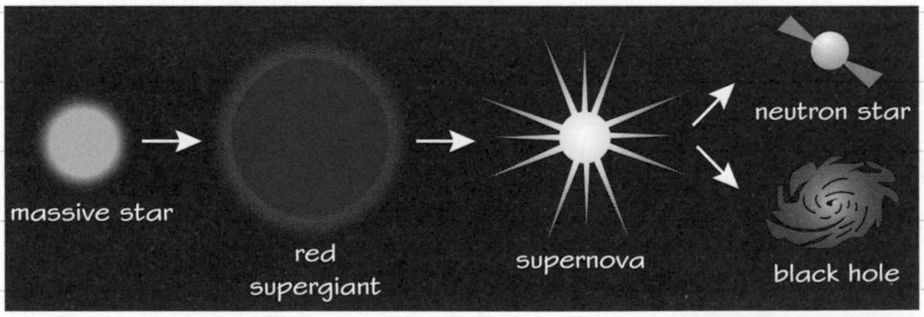

A star with a much higher mass than the Sun follows the same first stages of the life cycle, but each stage is shorter. When most of its hydrogen is used up it forms a **red supergiant**. At the end of this stage the star will explode as a **supernova**. If what remains after the explosion is less than four times the mass of the Sun it will be pulled together by gravity to form a very small, dense star called a **neutron star**. More massive remnants form **black holes**.

Worked example

Explain how the fate of a low-mass star is different from that of a high-mass star. **(6 marks)**

A low-mass star will move off the main sequence and expand to become a red giant before cooling and contracting to become a white dwarf. A high-mass star will move off the main sequence, become a red supergiant and explode in a supernova. The mass that is left in the core will become a very dense neutron star or a black hole.

Now try this

1 State the stages in the life cycle of
(a) low-mass stars **(4 marks)**
(b) high-mass stars. **(4 marks)**

2 Explain why the temperature of a white dwarf increases at first despite it losing thermal energy. **(2 marks)**

3 Explain the roles of gravity and thermal expansion in the stages of a star's life cycle. **(4 marks)**

Observing the Universe

Methods used to observe the Universe have changed over time. Originally, observations were made **from Earth**. Now there are **telescopes in space** to observe the Universe from above the Earth's atmosphere.

Modern telescopes

The first scientists explored the Universe by observing the **visible light** emitted by stars. They made more discoveries after telescopes were invented. Modern telescopes are very different to the early telescopes.

Development	Impact
greater magnifications	We can observe galaxies that are far away.
recording observations using photography or digital cameras	We can gather more data.
can be made with greater precision	We get clearer images.
telescopes that can detect other parts of the electromagnetic spectrum	We can observe objects in space that emit more radio waves, infrared, ultraviolet or X-rays than visible light.

Observing the Universe from Earth and from space

Some parts of the electromagnetic spectrum are absorbed by the Earth's atmosphere. Telescopes used to observe these wavelengths of radiation must be put on satellites in space.

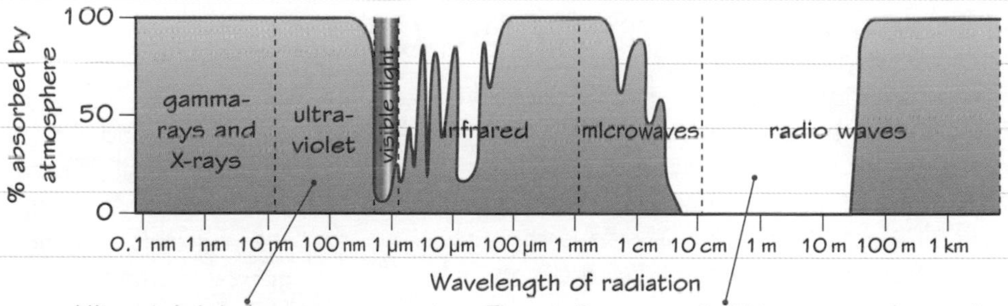

Ultraviolet telescopes need to be placed in orbit to detect ultraviolet radiation.

The radio waves in this range of wavelengths will reach us from space and be detected by radio telescopes on Earth. Longer wavelengths will not be detected.

Worked example

Suggest why the cosmic microwave background radiation was detected on Earth. **(2 marks)**

The longer wavelength microwaves that form the cosmic microwave background radiation are not absorbed by the Earth's atmosphere.

Use the diagram above to help with the answer.

Now try this

1 Explain why an X-ray telescope needs to be placed in orbit. **(2 marks)**
2 Explain why visible light telescopes can be based on Earth or in space. **(4 marks)**
3 Explain how developments in modern telescopes have improved our knowledge of the Universe. **(4 marks)**

Had a look ☐ Nearly there ☐ Nailed it! ☐

Extended response – Astronomy

There will be one or more 6 mark questions on your exam paper. For these questions, you will need to think scientifically and structure your answer logically, showing how the points you make are related to each other. You can revise the topics for this question, which is about **astronomy**, on pages 9 and 64–69.

Worked example

Astronomers collect data and use evidence to try to explain the origins of the Universe. For many years, there were two competing theories: the Steady State theory and the Big Bang theory.

Compare and contrast these theories.

Your answer should refer to the nature of any evidence which suggests that the Big Bang theory is the better theory. **(6 marks)**

The Big Bang theory states that the Universe started from a single, infinitely dense point in a massive explosion about 13.8 billion years ago and has been expanding and cooling ever since. The Steady State theory says that the Universe has always existed, and that it is also expanding.

The two key pieces of evidence for the Big Bang theory are the red-shift of galaxies and the cosmic microwave background radiation. Spectral lines in absorption spectra are shifted to the red end of the spectrum due to the stretching of light waves when galaxies move away from us. As the spectral lines from the majority of galaxies show red-shift, this suggests that the Universe is expanding. The cosmic microwave background radiation is the low-energy radiation or 'echo' of the Big Bang, at a temperature of 2.7 K, due to the cooling of gamma-rays that were emitted at the start of the Universe. The temperature of the cosmic microwave background radiation suggests that the Universe is the age that it has been predicted to be by the Big Bang theory.

The Big Bang theory is the more accepted theory because there is more evidence to support it as a theory for the Universe. Whilst the red-shift supports both theories, the cosmic microwave background radiation is strong evidence for the birth of the Universe in a Big Bang and not for a Steady State theory.

Decide what you are going to include in your answer, and plan the order of your answer before you start writing it. Think about the evidence you need to provide.

Command words: Compare and contrast

When you are asked to **compare and contrast**, you should describe the similarities and differences of both cases. You are not required to draw a conclusion.

When comparing and contrasting two theories you need to state how they are both similar and different. The main similarity and difference are stated here.

The two key pieces of evidence are stated and then described in more detail. The two pieces that need to be mentioned are red-shift and the cosmic microwave background radiation. There are other pieces of evidence, but stick with those mentioned on the specification.

Evidence is central to theories being accepted. You need to know what this evidence is so that you can explain why the Big Bang theory is the generally accepted one.

Now try this

Two stars are both in the main-sequence stage of their life. Describe what their possible fates could be and explain why these fates might be very different. **(6 marks)**

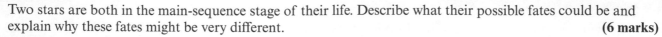

Work, energy and power

The **work** done by a force is the same as the **energy** transferred by the force. **Power** is the **rate of work** being done or the **rate at which energy is transferred**.

Work done and energy

Work done is the amount of energy transferred, and is measured in **joules (J)**.

The work done by a force is calculated using this formula:

work done (J) = force (N) × distance moved in the direction of the force (m)

$E = F \times d$

LEARN IT! IT'S NOT ON THE EQUATIONS LIST

E
F × d

You can measure the work done by recording the size of the force and the distance moved in the direction of the force. Having recorded these values, multiply them together to find the value for the work done.

Power

Power is the rate of doing work (how fast energy is transferred). Power is measured in **watts (W)**. 1 watt is 1 joule of energy being transferred every second.

$$\text{power (W)} = \frac{\text{work done (J)}}{\text{time taken (s)}}$$

$$P = \frac{E}{t}$$

LEARN IT! IT'S NOT ON THE EQUATIONS LIST

E
P × t

A hairdryer with a power of 1800 W transfers 1800 J each second.

Changing the energy of a system

The energy of a system can be changed by:

1. **work done through forces:** a body can be lifted through a vertical height by a force. This will increase the store of gravitational potential energy.
2. **electrical equipment:** a cell in a circuit provides a potential difference so that components in the circuit can transfer energy into other forms, such as thermal energy.
3. **heating a material:** supplying thermal energy to a system will increase the kinetic energy of the particles in the material. Its temperature will increase if it is not changing state.

Worked example

(a) Dan uses a force of 100 N to push a box across the floor. He pushes it for 3 m. Calculate the work done. **(3 marks)**

Work done = 100 N × 3 m = 300 J

(b) A kettle converts 360 000 J of electrical energy into thermal energy in 3 minutes whilst heating a mass of water. Calculate the power of the kettle. **(3 marks)**

Power = energy ÷ time

= 360 000 J ÷ (3 × 60 s) = 2000 W

Don't get work and power mixed up. Remember:

Work = energy transferred, measured in joules.

Power = *rate* of energy transfer, measured in watts.

Always make sure that energy is in J and time is in s before substituting to find an answer in W.

Now try this

1 Calculate how much work is done when 350 N pushes an object through 30 m along the floor. **(3 marks)**

2 Calculate how much energy an 1800 W hairdryer transfers when used for 8 minutes. **(3 marks)**

3 Calculate the power of a light bulb that transfers 3600 J of energy in one minute. **(3 marks)**

4 Calculate the temperature rise of a 36 kg mass of water when it is heated by a 4 kW heater for 46 minutes and 40 seconds. **(5 marks)**

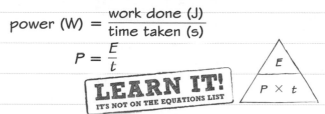

Had a look ☐ Nearly there ☐ Nailed it! ☐

Extended response – Energy and forces

There will be one or more 6 mark questions on your exam paper. For these questions, you will need to think scientifically and structure your answer logically, showing how the points you make are related to each other. You can revise the topics for this question, which is about **energy and forces**, on pages 17, 18, 21 and 71.

Worked example

You can find more information on efficiency on page 18.

Energy stores and energy transfers can be shown on energy transfer diagrams.

An energy transfer diagram is shown for a battery-operated fan.

Explain how this energy transfer diagram can be used to determine the efficiency of the fan.

Your answer should refer to the term efficiency, the equation for efficiency and a calculation. **(6 marks)**

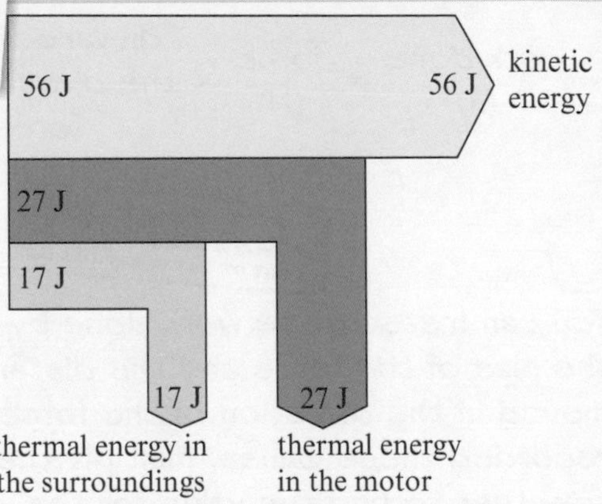

For the energy transfer diagram shown, the input is in the form of a chemical energy store, which is transferred to kinetic energy, sound energy and thermal energy.

For the energy transfer diagram shown, for every 100 J of energy transferred from the chemical energy store of the battery, there will be 56 J transferred to the kinetic energy, 17 J to the thermal energy store in the surroundings and 27 J to the thermal energy store in the motor.

Since a battery-operated fan is designed to transfer the chemical energy store to kinetic energy as a useful form, 56 J of the 100 J of chemical energy can be deemed to be useful, and 44 J can be assumed to be wasted.

The efficiency of a device is the proportion of the input energy that is usefully transferred and is calculated using the equation:

$$\text{Efficiency} = \frac{\text{(useful output energy)}}{\text{(total input energy)}} \times 100\%$$

For the battery-operated fan, the efficiency is 56 J ÷ 100 J × 100% which is an efficiency of 56%.

The answer starts by simply stating what the energy transfer diagram is showing in terms of energy stores and transfers.

This part of the answer explains which energy store is useful as an output energy store and which of the stores are wasted energy stores. It also states the amounts that are useful and wasted.

Command word: Explain

When you are asked to **explain** something, it is not enough just to state or describe it.

Your answer must contain some reasoning or justification of the points you make.

Your explanation can include mathematical explanations, if calculations are needed.

The term efficiency is explained, an equation for calculating the efficiency is provided and a value is calculated for the fan's efficiency.

The question expects you to talk about **energy stores** and **energy transfers**. There are eight types of energy store and 4 types of energy transfer, which you can read more about on page 17.

Now try this

Two athletes decide to compare themselves in terms of how powerful they are when trying to climb a hill. One of the athletes is a climber and the other is a runner.

Describe an experiment to see which of the two athletes is the most powerful.

Your answer should refer to the equations for calculating changes in gravitational potential energy and power. **(6 marks)**

Interacting forces

Pairs of forces can interact **at a distance** or by **direct contact**.

Non-contact forces

Forces can be exerted between objects without them being in contact with one another. There are three **non-contact forces** that you need to know about.

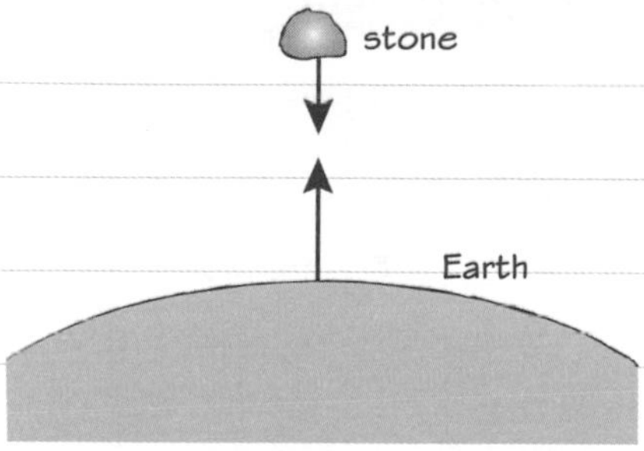

Gravity acts between all masses. It is always attractive.

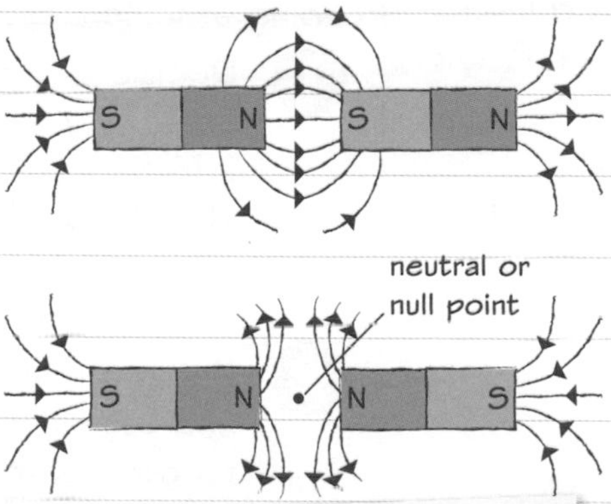

Magnetism acts between magnetic poles. Unlike poles attract and like poles repel.

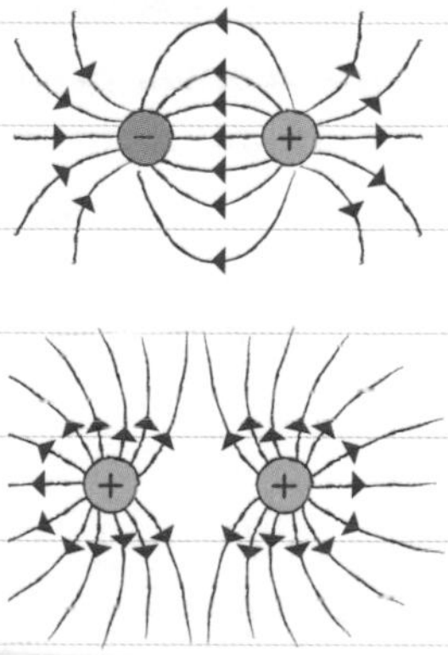

Electrostatic charges exert forces on each other at a distance. Unlike charges attract and like charges repel.

Contact forces

Forces can be exerted between objects due to them being in contact. In each case there is an interaction pair of forces that act in opposite directions. The forces can be represented by vectors.

The **normal contact force** acts upwards in opposition to the weight of the object. The interaction pair is the force of the ground on the box and the force of the box on the ground.

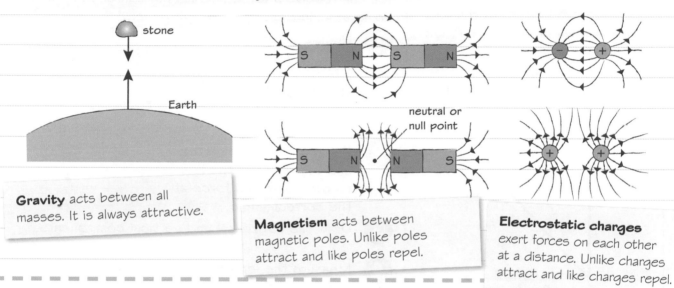

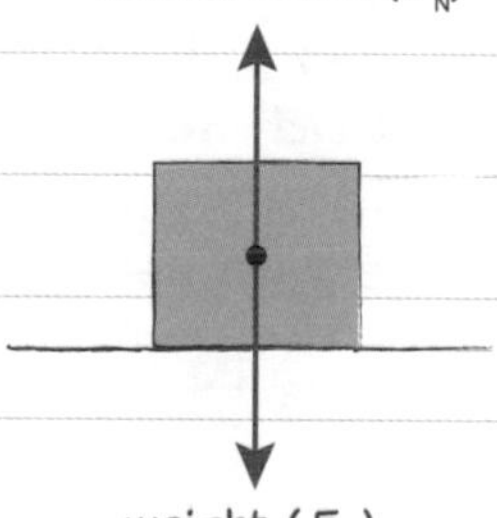

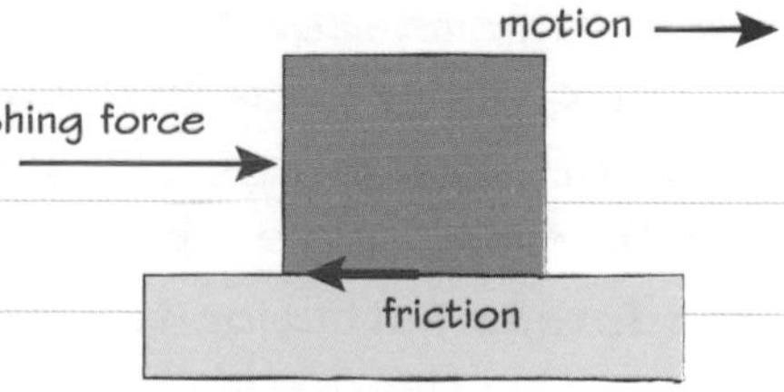

The **force of friction** acts in opposition to the pushing force that is trying to change its motion. Friction always acts to slow a moving object down. The interaction pair is the force of the object on the surface and the force of the surface on the object.

Worked example

An archer fires an arrow from a bow. Describe the contact and non-contact forces involved in firing the arrow. **(2 marks)**

Contact forces include the normal force acting on the arrow, friction as it moves through the air and the tension in the string. Non-contact forces include gravity once it has been released.

Other examples of contact forces include tension, upthrust and drag.

You can find more information about upthrust on page 119.

Now try this

1 State how gravity is different from magnetism and the electrostatic force. **(1 mark)**

2 State how the forces of friction and drag are similar. **(1 mark)**

3 Describe the contact and non-contact forces acting on a moving ship. **(3 marks)**

Free-body force diagrams

A **resultant force** can be resolved into its **horizontal** and **vertical** component forces. **Free-body force diagrams** can be drawn to show all the forces acting on a body.

Components of a force

Any force can be resolved into its **horizontal component** F_x and its **vertical component** F_y.

The horizontal and vertical components of motion do not affect one another – they are independent.

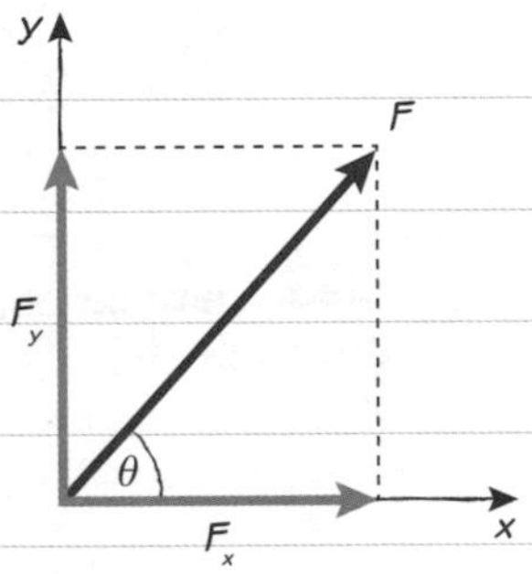

Resolving the forces by scale drawing

You can resolve a force into its horizontal components by doing a scale drawing on graph paper.

1. Decide on an appropriate scale. For example, for a force of 50 N acting at 60° to the horizontal use a scale of 1 cm represents 10 N.
2. Using a ruler, protractor and pencil, draw the line to represent the force at the correct angle.
3. Draw the horizontal and vertical components.
4. Measure the lengths and convert to a force using the same scale.

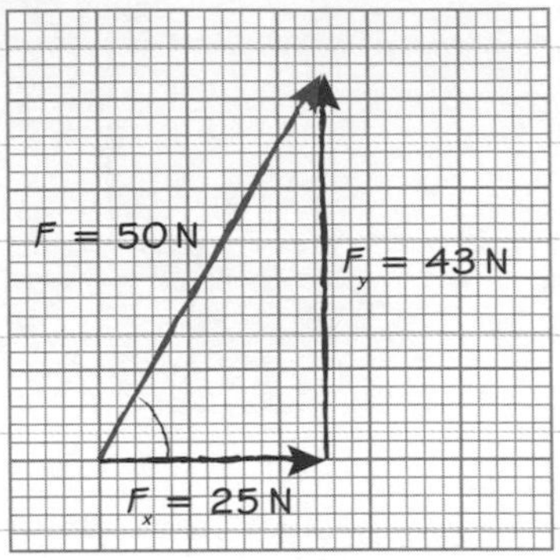

Free-body diagrams

Free-body diagrams show the forces acting on a single object. They need to contain:

1. **the body the forces are acting on** This can be simplified to a box or a dot.
2. **the forces acting on the body** This will depend on the complexity of the problem and the assumptions being made.

Free-body diagrams use arrows to show the **size** and **direction** of the forces involved. If these arrows **balance** then there is **no resultant force**. If the arrows do **not** balance then the body will be **accelerating**.

Free-body force diagrams

A force is a vector quantity, because it has a direction as well as a size. A **free-body force diagram** represents all the forces on a single body. Larger forces are shown using longer arrows.

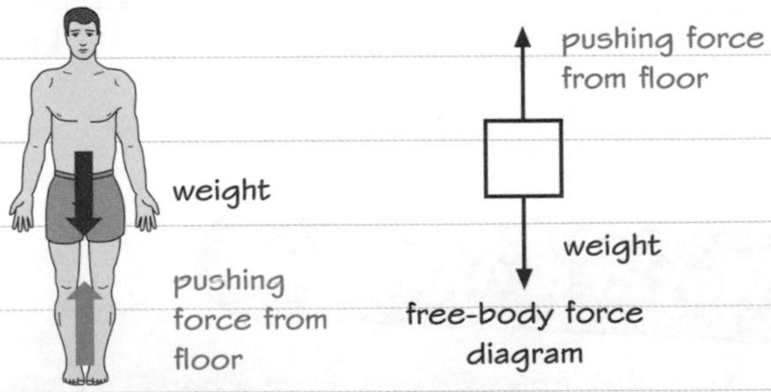

Worked example

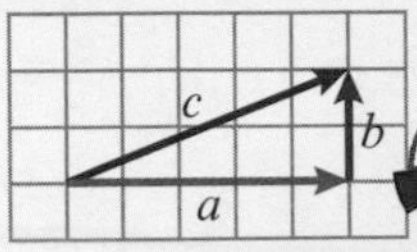

Each square on this 1 cm grid represents a force of 10 N. Use this to find:

(a) the resultant force **(2 marks)**

The length of the resultant force is 5.4 cm, so the resultant force is 54 N.

(b) the horizontal component of the force **(2 marks)**

The length of *b* is 5 cm, so the force is 50 N.

(c) the vertical component of the force. **(2 marks)**

The length of *a* is 2 cm, so the force is 20 N.

Now try this

1 Draw a 75 N force, acting at 40° to the horizontal, and resolve it into its horizontal and vertical components. **(4 marks)**

2 Draw a free-body diagram for a sky diver who has just jumped out of a plane from 3000 m. **(3 marks)**

Resultant forces

Resultant forces determine whether a body will be **stationary**, moving at a **constant speed** or **accelerating**.

Resultant forces

Forces are **vectors**, so they have a **size** and a **direction**. Both of these need to be taken into consideration when finding the resultant force.

The resultant force is the single force that would have the same effect as all of the other forces acting on the object. There can be many of these, but they always simplify to just one resultant force.

A resultant force of zero means a body is either stationary or moving at a constant speed.

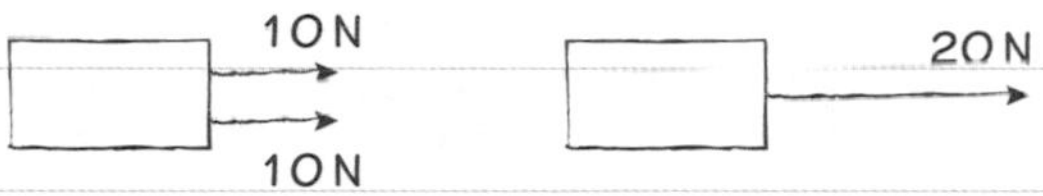

Forces acting in the same direction are added.

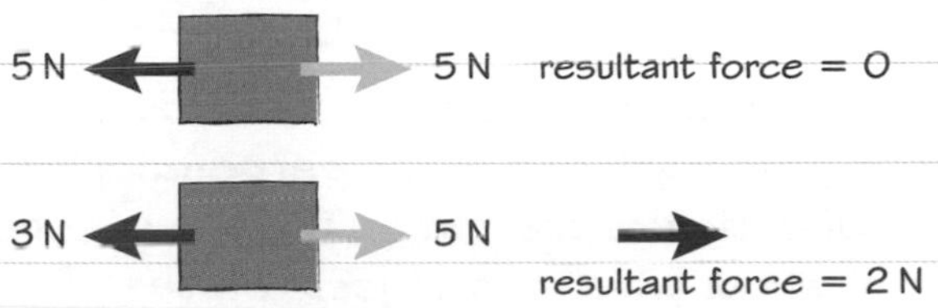

Forces acting in opposite directions are subtracted.

Resultant forces in 2D

Sometimes, forces act at right angles to each other. This means that they cannot simply be added or subtracted like parallel forces can be.

The resultant force can then be found from a scale drawing on graph paper.

When the forces are at right angles, you can check the answer using Pythagoras' theorem.

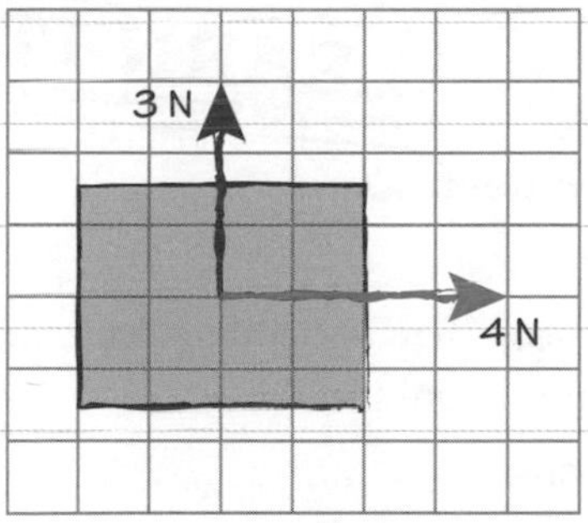

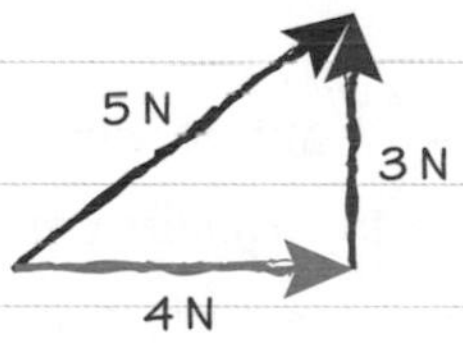

3 N acting up and 4 N acting to the right produce a 5 N resultant force in the direction shown.

Check using Pythagoras' theorem:

$$a^2 + b^2 = c^2$$

$$3^2 + 4^2 = 25$$

$$\text{resultant} = \sqrt{25} = 5\,\text{N}$$

angle of resultant $= \tan^{-1}\left(\frac{3}{4}\right) = 37°$ to the horizontal

Worked example

Work out the size and direction of the resultant force. **(5 marks)**

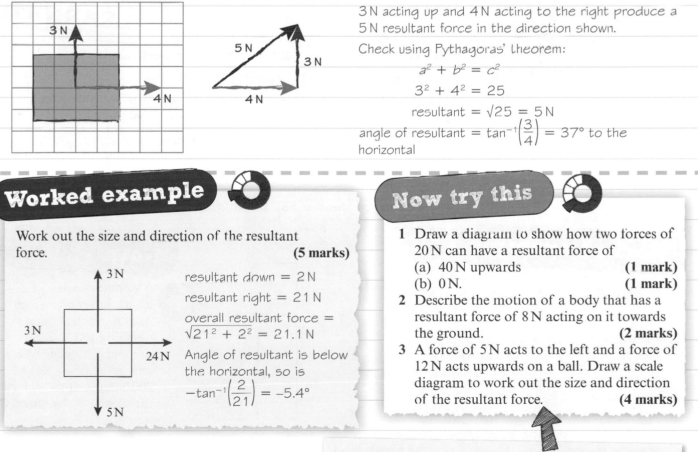

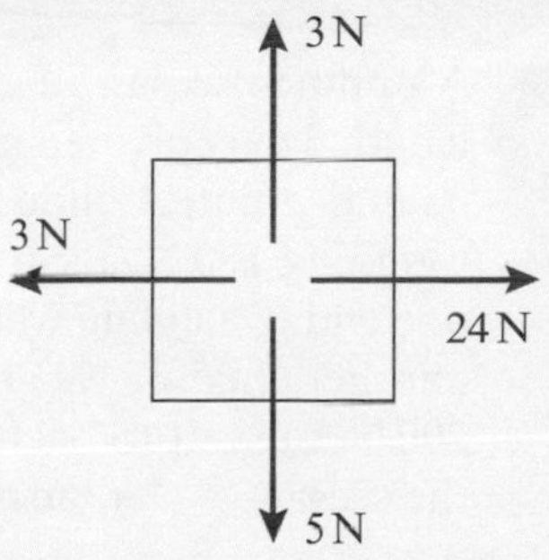

resultant down = 2 N

resultant right = 21 N

overall resultant force = $\sqrt{21^2 + 2^2} = 21.1\,\text{N}$

Angle of resultant is below the horizontal, so is

$-\tan^{-1}\left(\frac{2}{21}\right) = -5.4°$

Now try this

1 Draw a diagram to show how two forces of 20 N can have a resultant force of
   (a) 40 N upwards **(1 mark)**
   (b) 0 N. **(1 mark)**

2 Describe the motion of a body that has a resultant force of 8 N acting on it towards the ground. **(2 marks)**

3 A force of 5 N acts to the left and a force of 12 N acts upwards on a ball. Draw a scale diagram to work out the size and direction of the resultant force. **(4 marks)**

You can check your answer using Pythagoras' theorem.

Moments

Forces which act at a **distance** from a **pivot** can cause a **turning effect** or rotation. This is known as a **moment**.

Moments

The size of the moment, or turning effect, is given by the equation:

moment (newton-metres, Nm) = force (N) × distance normal to the direction of the force (m)

LEARN IT! IT'S NOT ON THE EQUATIONS LIST

The distance must be measured from the pivot normal (at right angles, or **perpendicular**) to the direction of the force. The pivot is the point about which the object rotates.

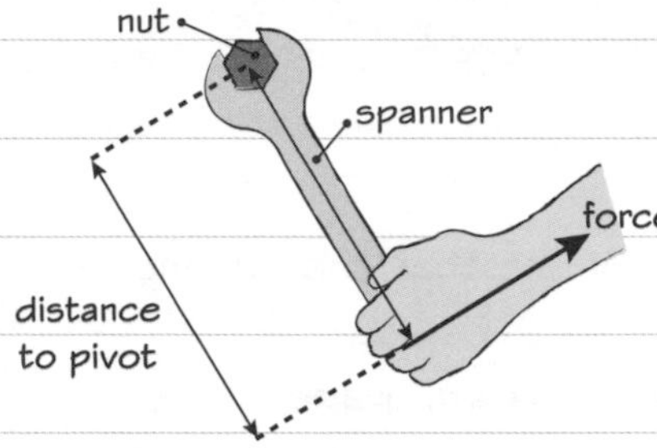

Worked example

Two children are sitting on a seesaw.

(a) Calculate the anticlockwise moment. **(2 marks)**

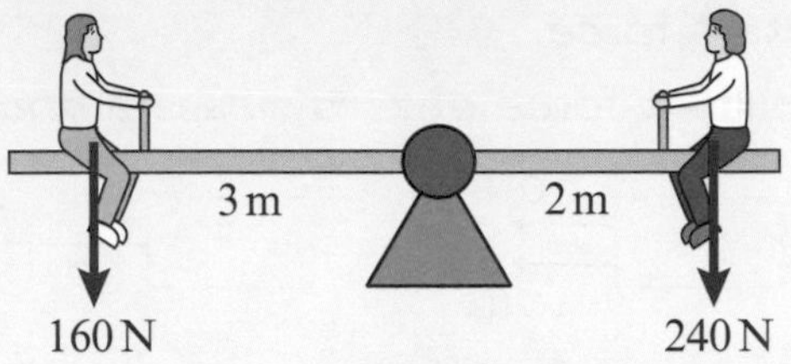

= 160 N × 3 m = 480 Nm

(b) Calculate the clockwise moment. **(2 marks)**

= 240 N × 2 m = 480 Nm

(c) Explain why the seesaw doesn't move. **(2 marks)**

The anticlockwise and clockwise moments about the pivot are equal so the seesaw is balanced and does not move.

The principle of moments

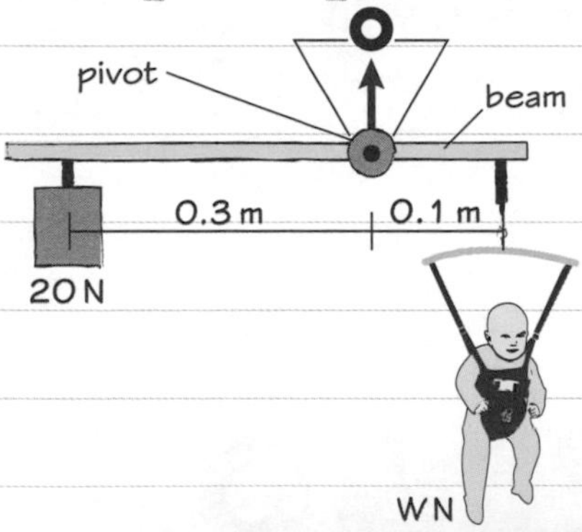

LEARN IT! IT'S NOT ON THE EQUATIONS LIST

sum of the clockwise moments = sum of the anticlockwise moments

for rotational forces in equilibrium.

The diagram shows a common arrangement for weighing fruit and vegetables, as well as babies. The beam is balanced so:

anticlockwise moment = clockwise moment

20 N × 0.3 m = *W*N × 0.1 m

W = 6 Nm/0.1 m = 60 N

The weight of the bar has been ignored in this calculation.

Worked example

A ladder is resting against a wall. The weight of the ladder is 20 N.Calculate the force that the wall applies to the top of the ladder. **(5 marks)**

Take moments from the point at which the floor meets the wall.

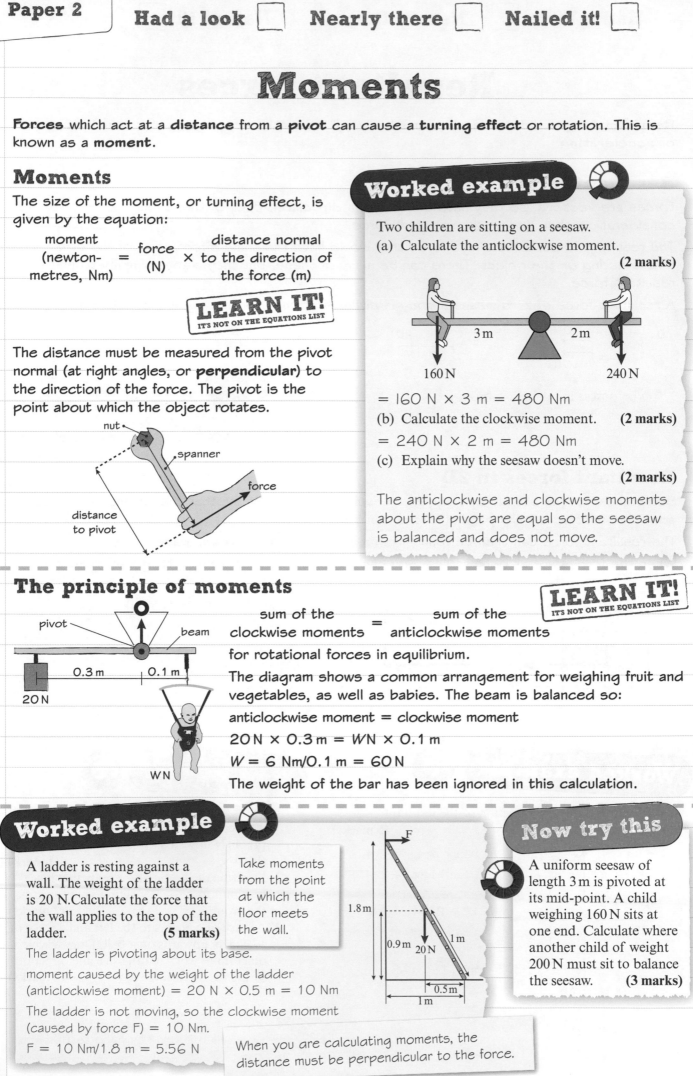

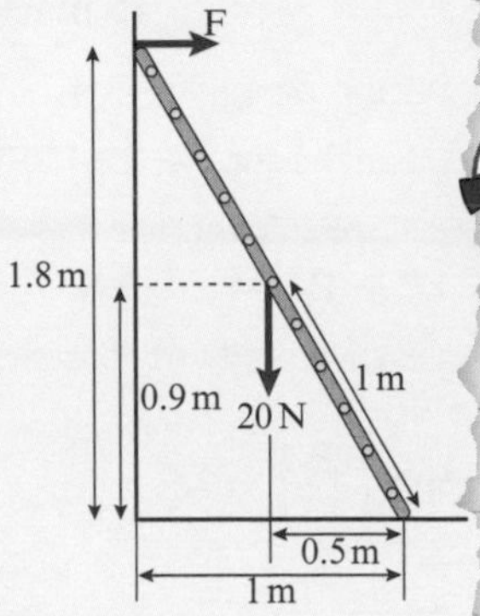

The ladder is pivoting about its base.

moment caused by the weight of the ladder (anticlockwise moment) = 20 N × 0.5 m = 10 Nm

The ladder is not moving, so the clockwise moment (caused by force F) = 10 Nm.

F = 10 Nm/1.8 m = 5.56 N

When you are calculating moments, the distance must be perpendicular to the force.

Now try this

A uniform seesaw of length 3 m is pivoted at its mid-point. A child weighing 160 N sits at one end. Calculate where another child of weight 200 N must sit to balance the seesaw. **(3 marks)**

Levers and gears

Levers and **gears** can transmit the **rotational effect** of forces.

Levers

A **lever** can be used to cause rotation. The nature of this rotation depends on the position of:

1. the **input force** – the force provided by the user of the lever
2. the **output force** – the force that results from the input force
3. the **fulcrum** – the turning point about which both forces act.

For different levers, the input force, output force and fulcrum are in different positions relative to each other.

class 1 lever

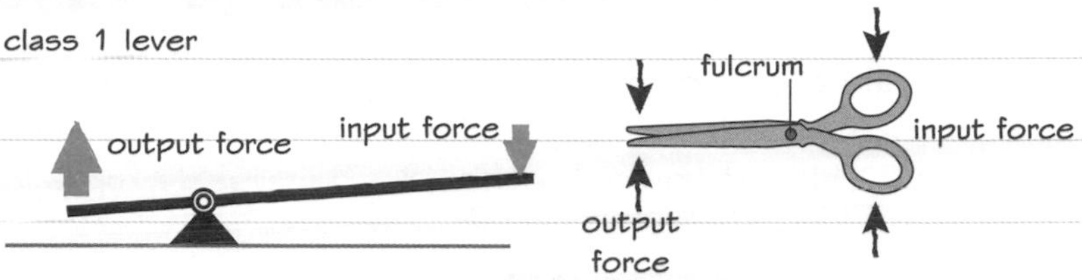

class 2 lever

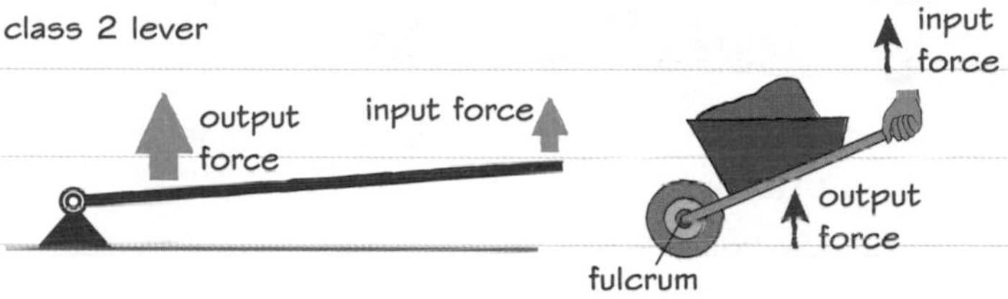

class 3 lever

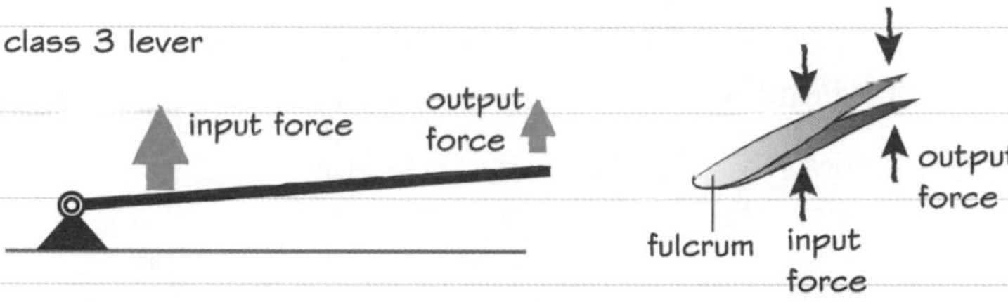

Gears

Gears are toothed wheels that are meshed together to transmit rotational force and motion. This is a low gear.

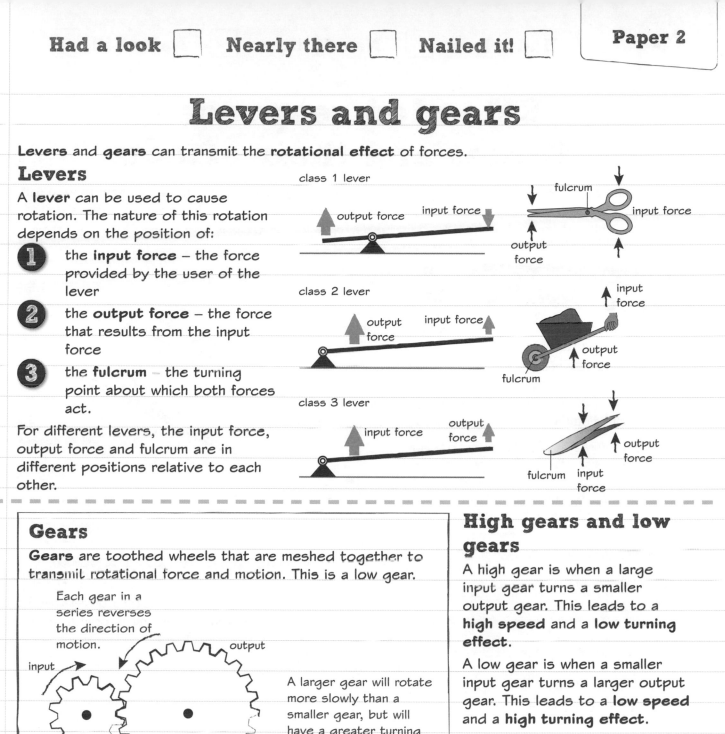

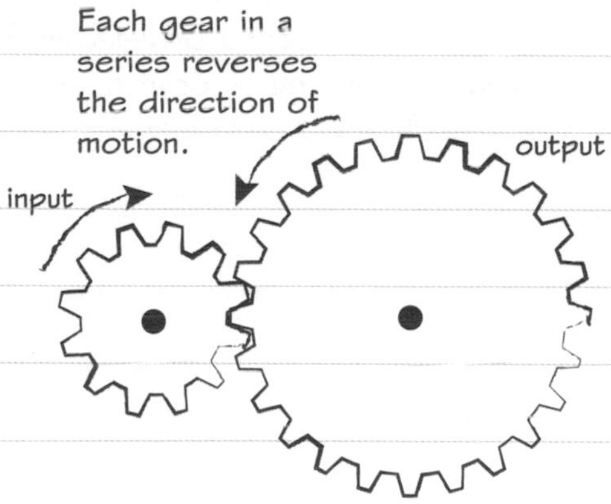

A larger gear will rotate more slowly than a smaller gear, but will have a greater turning effect. A low gear leads to a low speed and a high turning effect.

High gears and low gears

A high gear is when a large input gear turns a smaller output gear. This leads to a **high speed** and a **low turning effect**.

A low gear is when a smaller input gear turns a larger output gear. This leads to a **low speed** and a **high turning effect**.

Worked example

Describe the common features of gears and levers in terms of how they work. **(2 marks)**

Gear and levers both transmit the rotational effect of a force and they can alter the size of the output force compared with the input force.

You need to know that both gears and levers can cause rotation and can alter the size of the output force.

Now try this

Look at page 76 on moments. Consider the turning moment and describe how this affects the relative sizes of the forces.

1 Give one example of an
(a) class 1 lever **(1 mark)**
(b) class 2 lever **(1 mark)**
(c) class 3 lever. **(1 mark)**

2 Explain why a bicycle or a car needs different gears. **(3 marks)**

3 Explain why a heavy rock is easier to lift with a class 1 lever than without. **(4 marks)**

Extended response – Forces and their effects

There will be one or more 6 mark questions on your exam paper. For these questions, you will need to think scientifically and structure your answer logically, showing how the points you make are related to each other. You can revise the topics for this question, which is about **forces and their effects**, on pages 73–77.

Worked example

The free-body diagram shows the forces acting on a skier. The skier is moving down the slope.

Explain what the possible motion of the skier could be with reference to the diagram.

Your answer should refer to the size, nature and direction of the forces acting on the skier. **(6 marks)**

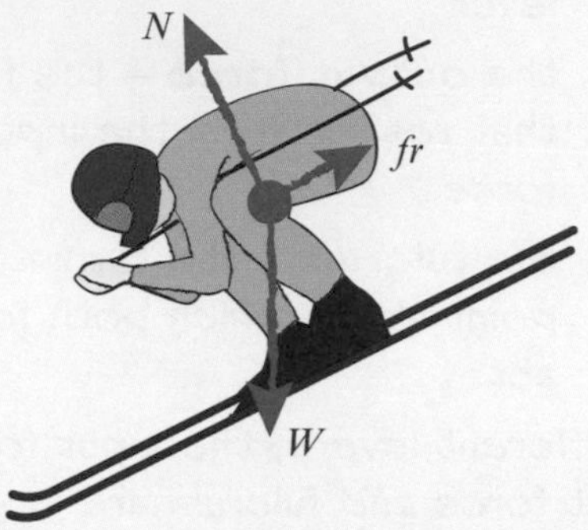

There are three forces shown on the diagram – the weight of the skier acting vertically downwards, the normal contact force of the skier, acting at right angles to the slope, and the force of friction acting up the slope.

The weight is caused by the gravitational force, a non-contact force, acting on the skier's mass. The reaction force is a contact force acting perpendicular to the slope due to the slope pushing upwards on the skier. The force of friction is a contact force caused when the skis are in contact with the snow or ice. This force will act parallel to the slope and up the slope in the opposite direction to the skier's motion.

The skier will accelerate down the slope if the component of his weight acting down the slope is greater than the force of friction acting up the slope. If the component of the weight acting down the slope is less than the friction, the skier will slow down. If these two forces are balanced then the skier will either be stationary or will move at a constant speed.

The three forces are identified and stated. Their directions are also provided.

Command word: Explain

When you are asked to **explain** something, it is not enough to just state or describe something. Your answer must contain some reasoning or justification of the points you make.

Having stated the names of the three forces, a description is then given as to how they have been caused. Reference to contact and non-contact forces is also provided in the answer.

Having stated the nature and direction of the three forces, the motion of the skier can be explained by referring to the size of any balanced or unbalanced forces.

A body will be stationary or move at a constant speed if the forces acting on it are balanced. Since the question states that the skier is moving down the slope, do not state that the skier is stationary in the answer or make it clear that you have discounted the possibility because of what the question states. The acceleration of any body, including the skier, is explained by Newton's second law of motion. You can read about this on page 8.

Now try this

Describe what can happen when two objects interact with one another. Your answer should refer to contact and non-contact forces. **(6 marks)**

Circuit symbols

Electric circuit diagrams are drawn using agreed symbols and conventions that can be understood by everybody across the world.

Circuit symbols

You need to know the symbols for these components.

Component	Symbol	Purpose
cell	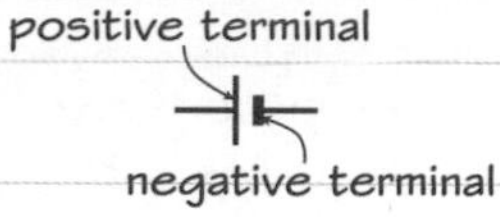	provides a potential difference
battery		provides a potential difference
switches		allows the current flow to be switched on or off
voltmeter	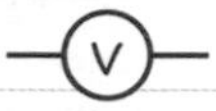	measures potential difference across a component
ammeter		measures the current flowing through a component
fixed resistor		provides a fixed resistance to the flow of current
variable resistor		provides a variable (changeable) resistance
filament lamp		converts electrical energy to light energy as a useful form
motor	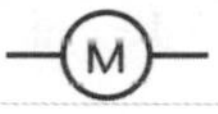	converts electrical energy to kinetic energy as a useful form
diode	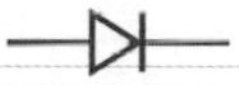	allows current to flow in one direction only
thermistor		resistance decreases when the temperature increases.
LDR	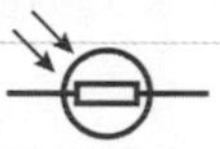	resistance decreases when the light intensity increases.
LED	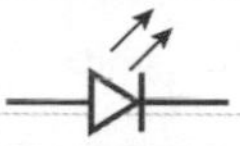	a diode that gives out light when current flows through it

Worked example

Name four components that are commonly used to change or control the amount of resistance in a circuit. **(2 marks)**

thermistor, LDR, fixed resistor, variable resistor

2 marks for all correct, 1 mark for 3 correct and zero marks for less than 3 correct.

Worked example

Describe how an LED is different from a filament lamp. **(1 mark)**

An LED will only allow electrical current to pass through it in one direction.

Now try this

1 State which components above are output devices. **(3 marks)**

2 State how ammeters and voltmeters should be arranged with components in circuits. **(2 marks)**
3 Design a circuit to turn on a lamp when it goes dark. **(3 marks)**

Series and parallel circuits

Components in circuits can be arranged in **series** or **parallel**. The rules for current and potential difference in these two types of circuits are different.

Series circuits

A **series circuit** contains just one loop, around which an electric current can flow.

Electrons do not flow through any of the voltmeters.

Electrons flow clockwise in this circuit, through the cell, wires, three ammeters and two resistors.

Ammeters are always connected in series with components.

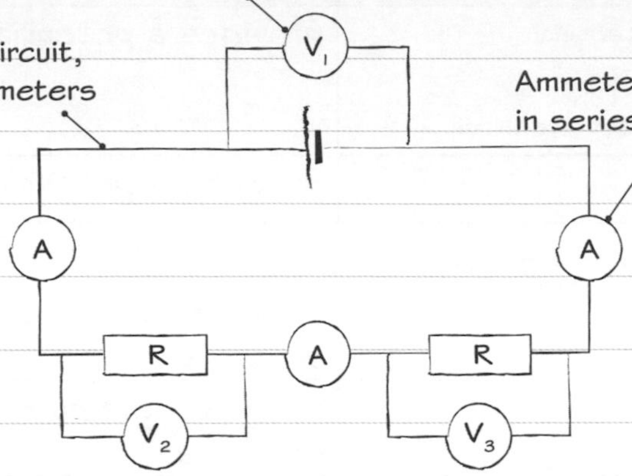

Ammeters have a very low resistance so the current can flow through them and be measured accurately.

The size of the current in a series circuit is the same at every point in the circuit. All three ammeters in this circuit will show the same value for the current.

The potential difference across components that are arranged in series must add up to give the cell voltage, so $V_1 = V_2 + V_3$.

Voltmeters have a very high resistance so that no current will flow through them.

Parallel circuits

A **parallel circuit** contains more than one loop and the current will split up or recombine at the junctions.

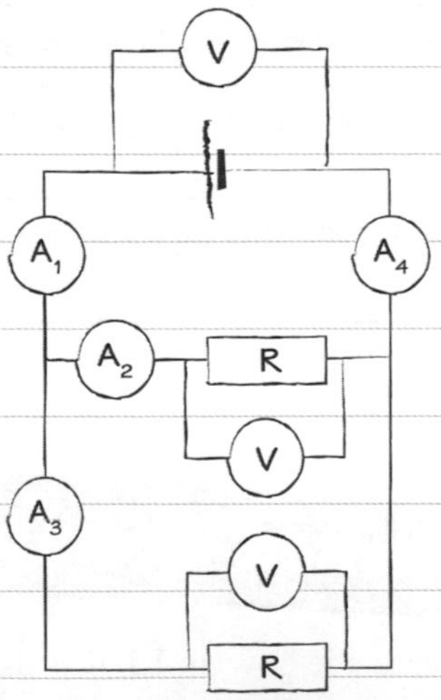

The potential difference across the components in each branch of a parallel circuit must add to give the cell voltage. The voltmeter readings across the two resistors will both have the same value as the potential difference across the cell.

The sum of the currents in each of the branches must equal the current leaving the cell. For this example, $A_1 = A_4 = A_2 + A_3$.

Worked example

What are the missing values in this parallel circuit if the bulbs are identical? **(2 marks)**

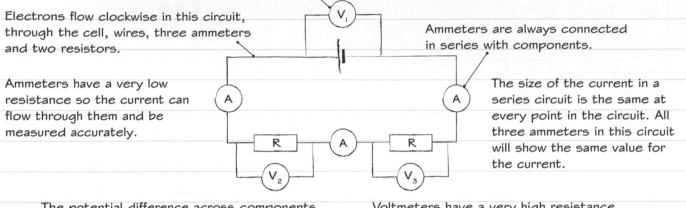

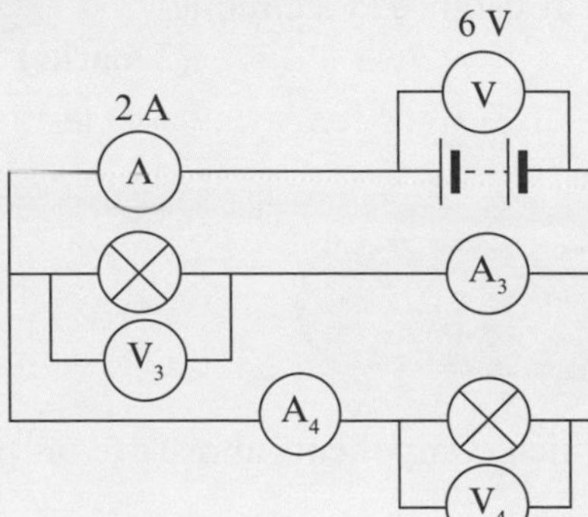

Both the ammeter readings A_3 and A_4 = 1 A. The voltmeter readings V_3 and V_4 will both be 6 V.

Now try this

1 Describe the differences for current and potential difference in series and parallel circuits. **(4 marks)**

2 Explain why car lights are connected in parallel and not in series. **(3 marks)**

3 Explain why a cell goes flat more quickly when lamps are arranged in parallel. **(4 marks)**

Current and charge

An **electric current** is the **rate of flow of charge.** In a metal, electric current is the flow of electrons.

Charge and current

The size of a current is a measure of how much charge flows past a point each second. It is the rate of flow of charge. The unit of charge is the **coulomb** (C). One ampere, or amp (A), is one coulomb of charge per second. You can calculate charge using the equation:

Charge (C) = current (A) × time (s)

$Q = I \times t$

LEARN IT! IT'S NOT ON THE EQUATIONS LIST

Watch out! Don't get the units and quantities confused. The units have sensible abbreviations (C for coulombs, A for amps). The symbols for the quantities are not as easy to remember (Q stands for charge, I for current).

You also need to know how to rearrange the equation using the triangle shown so that you can calculate Q, I and t using values provided in questions.

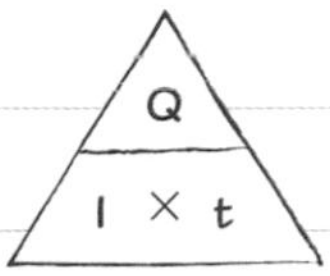

Measuring current

Electric current will flow in a closed circuit when there is a source of potential difference. To measure the size of the current flowing through a component, an ammeter is connected in series with the component.

The current flowing is the same at all points in a series circuit. In this example, the current is 0.5 A.

The cell is the source of potential difference.

Conventional current flows this way, from + to – in a circuit.

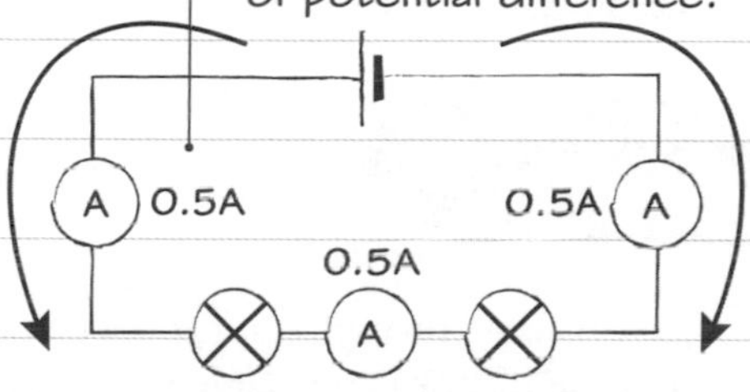

Electrons flow this way, from – to +.

The three ammeters are connected in series with the two filament lamps.

Worked example

(a) Describe how ammeters should be connected with components in circuits. **(1 mark)**

Ammeters should always be arranged in series with the component to be measured.

(b) Describe what happens to the size of the current in a series circuit. **(1 mark)**

The current is the same value at any point in a series circuit.

Worked example

(a) A current of 1.5 A flows for 2 minutes. Calculate how much charge flows in this time. **(3 marks)**

$Q = I \times t$

$= 1.5\,A \times 120\,s = 180\,C$

(b) A charge of 1200 C flows through a filament lamp for 4 minutes. Calculate the average current in the filament lamp. **(3 marks)**

$I = Q \div t$

$= 1200\,C \div 240\,s$

$= 5\,A$

Now try this

1 State the units of current and charge. **(2 marks)**

2 Calculate the charge that flows in one hour when there is a current of 0.25 A in a circuit. **(3 marks)**

3 A charge of 3×10^4 C is transferred by a current of 250 mA. Calculate the length of time the current flows. **(4 marks)**

Energy and charge

Energy, **charge** and **potential difference** are closely related when dealing with electrical circuits.

Energy, charge and potential difference

Energy, charge and potential difference are related by the equation:

energy transferred (J) = charge moved (C) × potential difference (V)

$E = Q \times V$

LEARN IT! IT'S NOT ON THE EQUATIONS LIST

Maths skills Cover up the quantity you want to find with your finger. The position of the other two quantities tells you the formula.

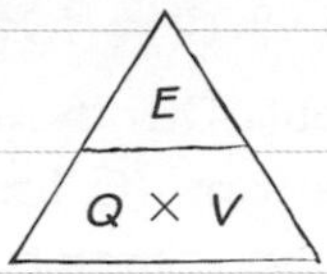

Calculate the charge, Q, from the current reading on the ammeter using the equation $Q = I \times t$ (see page 81).

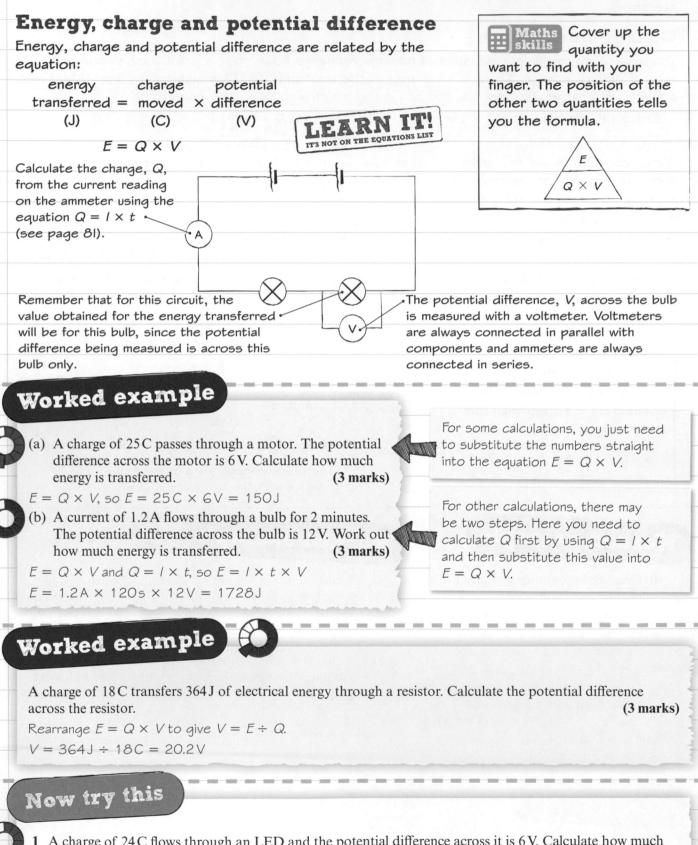

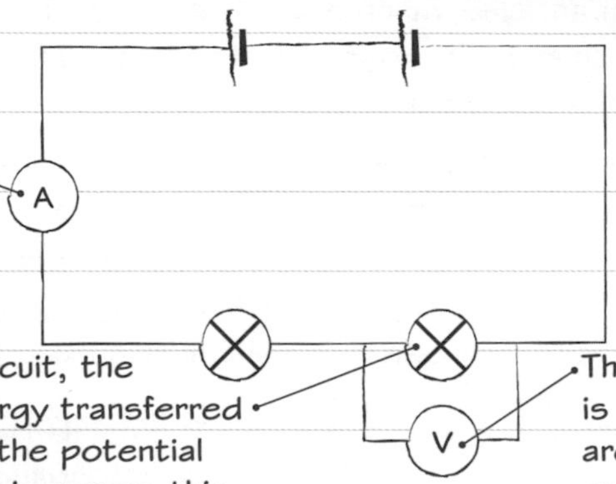

Remember that for this circuit, the value obtained for the energy transferred will be for this bulb, since the potential difference being measured is across this bulb only.

The potential difference, V, across the bulb is measured with a voltmeter. Voltmeters are always connected in parallel with components and ammeters are always connected in series.

Worked example

(a) A charge of 25 C passes through a motor. The potential difference across the motor is 6 V. Calculate how much energy is transferred. **(3 marks)**

$E = Q \times V$, so $E = 25\,C \times 6\,V = 150\,J$

For some calculations, you just need to substitute the numbers straight into the equation $E = Q \times V$.

(b) A current of 1.2 A flows through a bulb for 2 minutes. The potential difference across the bulb is 12 V. Work out how much energy is transferred. **(3 marks)**

$E = Q \times V$ and $Q = I \times t$, so $E = I \times t \times V$

$E = 1.2\,A \times 120\,s \times 12\,V = 1728\,J$

For other calculations, there may be two steps. Here you need to calculate Q first by using $Q = I \times t$ and then substitute this value into $E = Q \times V$.

Worked example

A charge of 18 C transfers 364 J of electrical energy through a resistor. Calculate the potential difference across the resistor. **(3 marks)**

Rearrange $E = Q \times V$ to give $V = E \div Q$.

$V = 364\,J \div 18\,C = 20.2\,V$

Now try this

1 A charge of 24 C flows through an LED and the potential difference across it is 6 V. Calculate how much energy is transferred. **(3 marks)**

2 Explain why the volt, V, can also be described as a 'joule per coulomb'. **(3 marks)**

3 The potential difference across a resistor is 18 V and it transfers 500 J in 4 minutes. Calculate the current passing through the resistor. **(4 marks)**

Ohm's law

Ohm's law states how the **current** through a component relates to its **resistance** and the **potential difference** across it.

Resistance

The **resistance** of a component is a way of measuring how hard it is for electricity to flow through it. The units for resistance are **ohms (Ω)**.

The resistance of a whole circuit depends on the resistances of the different components in the circuit. The higher the total resistance, the smaller the current.

resistance UP, current DOWN

The resistance of a circuit can be changed by putting different **resistors** into the circuit, or by using a **variable resistor**. The resistance of a variable resistor can be changed using a slider or knob.

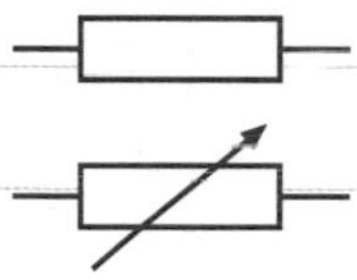

Ohm's law

Ohm's law states that the size of the current, I, flowing through a component of resistance, R, is directly proportional to the potential difference, V, across the component at constant temperature.

The equation for Ohm's law is:

potential difference (V) = current (A) × resistance (Ω)

$$V = I \times R$$

LEARN IT! IT'S NOT ON THE EQUATIONS LIST

Components that obey Ohm's law are said to be ohmic conductors whereas those that do not obey Ohm's law are non-ohmic. Examples of both of these can be found on page 85.

Worked example

(a) Resistor A has a current of 3 A flowing through it when the potential difference across it is 15 V. What is the size of its resistance? **(3 marks)**

From $V = I \times R$ you obtain $R = V / I$

Substituting gives $R = 15\,V / 3\,A = 5\,\Omega$

(b) Sketch a circuit that could be used to plot current against potential difference for this resistor. **(4 marks)**

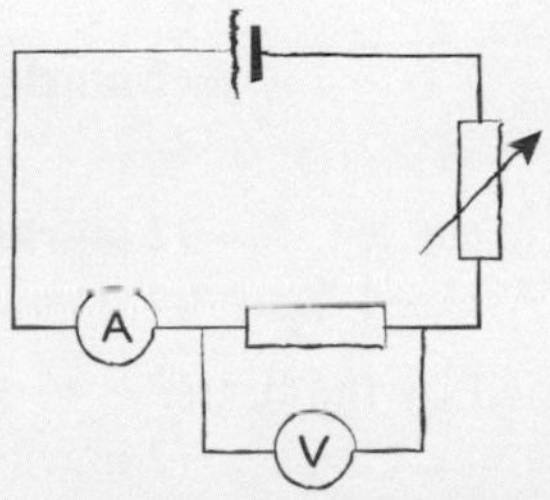

The variable resistor can be used to collect numerous values for I and V so that an I–V graph can be plotted.

(c) Resistor A is then replaced with resistor B. A graph of current against potential difference is plotted.

Explain what the graph tells you about the two resistors. **(2 marks)**

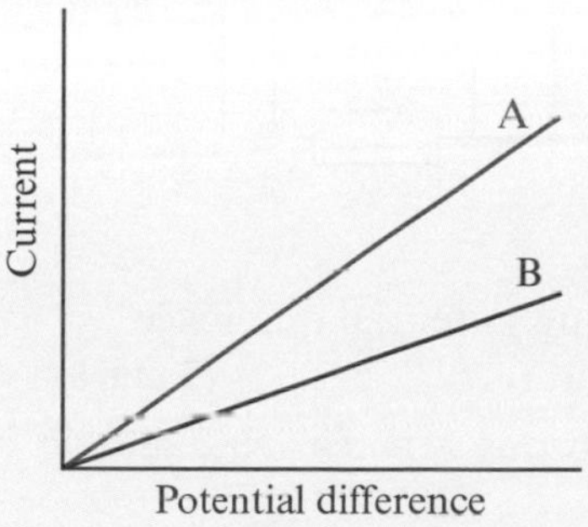

The steeper line shows that resistor A has a lower resistance than resistor B.

Both resistors are ohmic because the graphs are straight lines.

Now try this

1 A current of 3.2 A flows through a lamp of resistance 18 Ω. Calculate the size of the potential difference across the lamp. **(3 marks)**

2 The potential difference across a resistor is 28 V when the current flowing through it is 0.4 A. Calculate the resistance of the resistor. **(3 marks)**

Resistors

The arrangement of resistors in **series** or **parallel** determines whether the **resistance** in a circuit **increases** or **decreases**.

Resistors in series

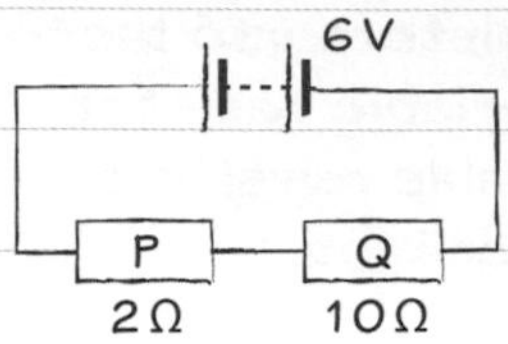

Calculating series resistance

For resistors arranged in series:

1. The current through each resistor is the same value.
2. The sum of the voltages across the resistors must add up to equal the cell voltage.

The total resistance in this series circuit is 12 Ω. The total potential difference from the battery is 6 V. Using $I = V \div R$ the current, I, flowing in resistors P and Q, is 6 V ÷ 12 Ω = 0.5 A.

Resistors in parallel

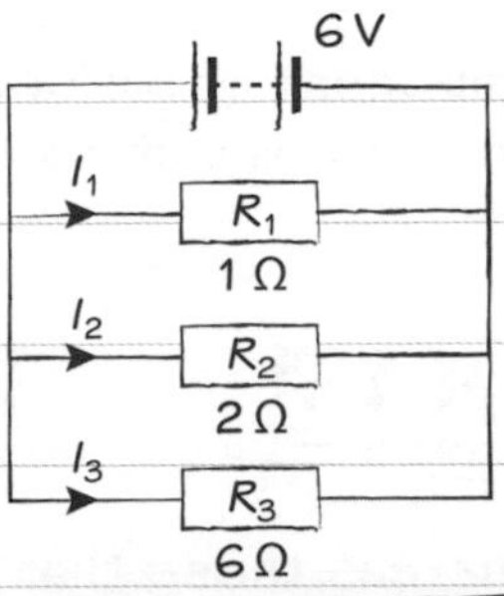

Calculating parallel resistance

For resistors arranged in parallel:

1. The total current leaving the battery is equal to the sum of the current flowing in the separate branches.
2. The potential difference across the resistors in each branch is equal to the potential difference of the battery.

Worked example

In the series circuit in the diagram, resistor X is known to have a resistance of 16 Ω.

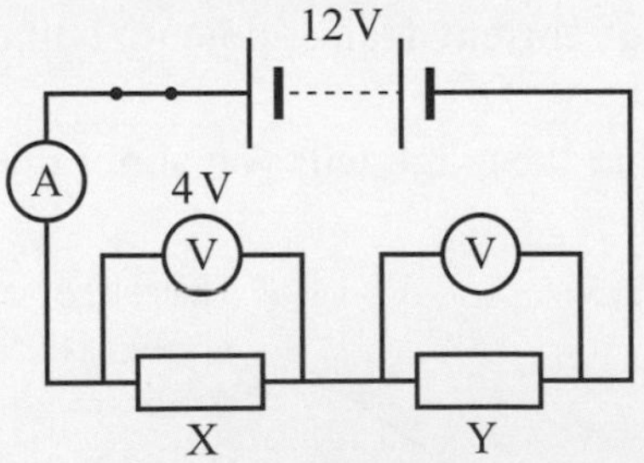

(a) Calculate the missing potential difference value across resistor Y. **(2 marks)**

8 V, since the values across X and Y must add up to 12 V.

(b) Calculate the current flowing through resistor X. **(3 marks)**

$I = V \div R$ = 4 V ÷ 16 Ω = 0.25 A

Worked example

Two resistors are arranged in parallel as shown in the circuit below. Calculate:

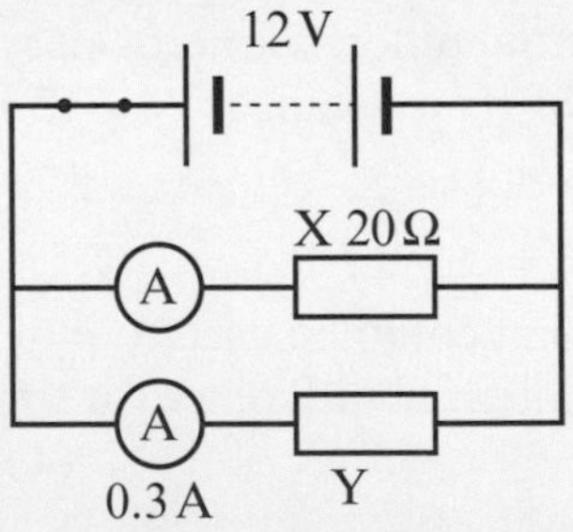

(a) the resistance of Y **(2 marks)**

$R = V \div I$ = 12 V ÷ 0.3 A = 40 Ω

(b) the current through resistor X **(2 marks)**

$I = V \div R$ = 12 V ÷ 20 Ω = 0.6 A

(c) the total current supplied by the battery. **(2 marks)**

total current = 0.3 A + 0.6 A = 0.9 A

Now try this

Two resistors, each of resistance 60 ohms, are arranged in series and connected to a 12 V cell. Calculate the size of the current flowing in the circuit. **(3 marks)**

I–V graphs

An **I–V graph** shows how the **current** flowing through a component varies as the **potential difference** across it varies. You need to know the characteristics of these I–V graphs.

1 Fixed resistor

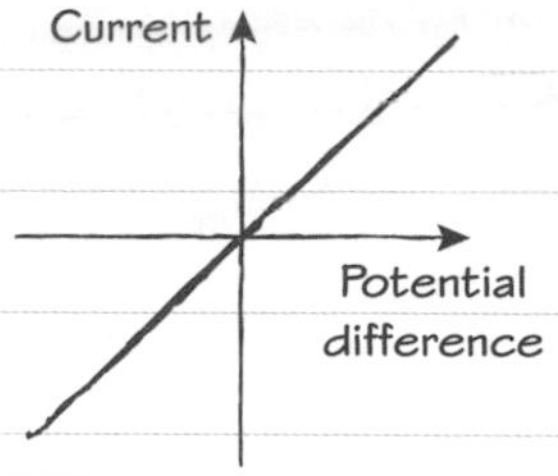

The temperature remains constant so the resistance remains constant. The slope or gradient is constant and the line remains straight throughout.

2 Filament lamp

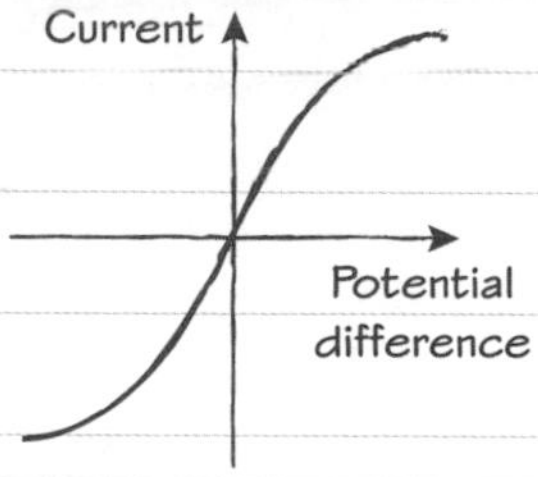

As the potential difference increases, the filament gets hotter, and atomic vibration increases. This leads to greater resistance. The slope or gradient decreases as the potential difference increases since resistance is increasing.

3 Diode

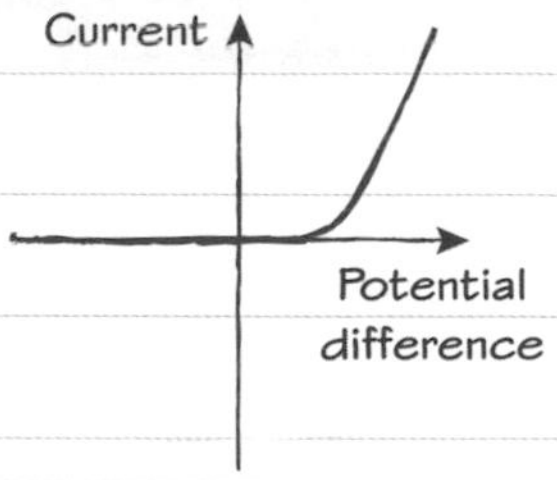

The current only flows in one direction. There is a **threshold** in the forward direction, which is why the graph is flat initially. A diode behaves like a fixed resistor – the resistance does not change. A diode has a very high resistance in the reverse direction.

Drawing an I–V graph

You can use a circuit like this to collect data about current and potential difference.

- ✓ When the switch is closed you can read current from the ammeter and potential difference from the voltmeter.
- ✓ Varying the value of the variable resistor allows you to record the current for different potential differences.
- ✓ You can reverse the cell to obtain negative values for potential difference.

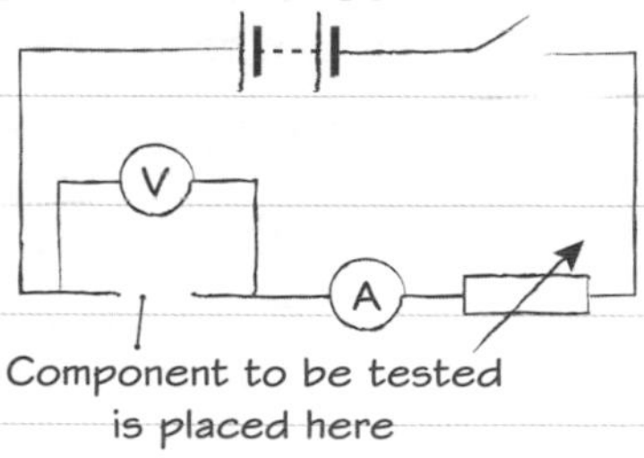

Worked example

A student is testing the resistance of a component. She draws an I–V graph.

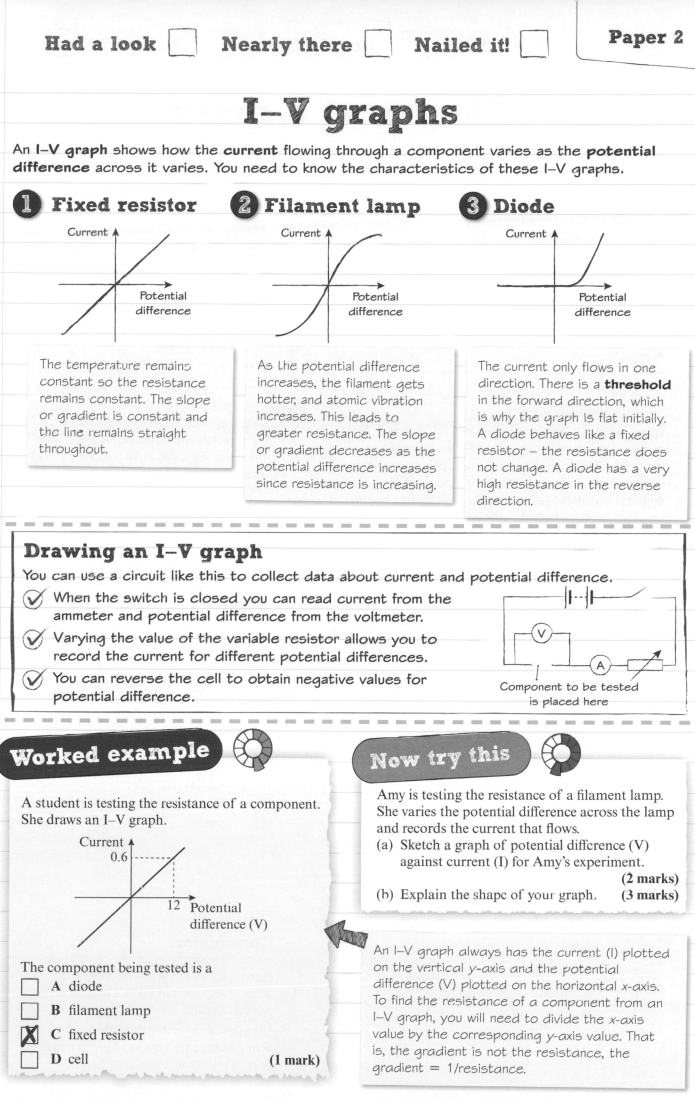

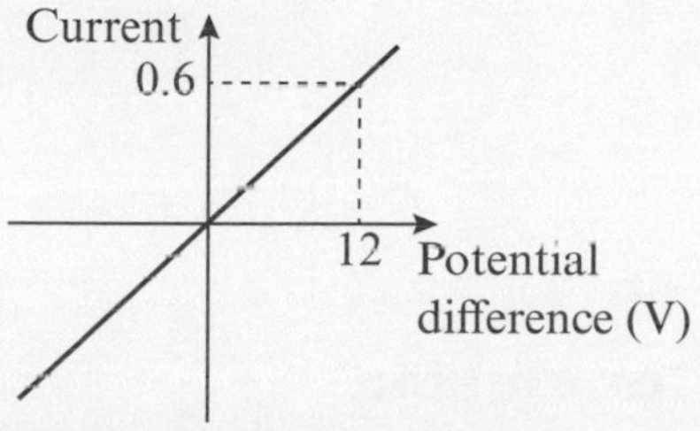

The component being tested is a

- [ ] **A** diode
- [ ] **B** filament lamp
- [x] **C** fixed resistor
- [ ] **D** cell

(1 mark)

An I–V graph always has the current (I) plotted on the vertical *y*-axis and the potential difference (V) plotted on the horizontal *x*-axis. To find the resistance of a component from an I–V graph, you will need to divide the *x*-axis value by the corresponding *y*-axis value. That is, the gradient is not the resistance, the gradient = 1/resistance.

Now try this

Amy is testing the resistance of a filament lamp. She varies the potential difference across the lamp and records the current that flows.

(a) Sketch a graph of potential difference (V) against current (I) for Amy's experiment. **(2 marks)**

(b) Explain the shape of your graph. **(3 marks)**

Electrical circuits

Practical skills You can investigate the relationship between **potential difference, current** and **resistance** for a filament lamp and a resistor by arranging them in series and parallel circuits.

Core practical

Aim

To determine the relationship between potential difference, current and resistance for a filament lamp and a resistor when arranged in series and parallel.

Apparatus

cell or battery, ammeters, voltmeters, filament lamps, resistors, connecting wires

Method 1: Investigating V, I and R

Set up the circuit so that the current through the filament lamp and the potential difference across it can be measured. Adjust the variable resistor setting to obtain a number of readings for current and potential difference. Plot a graph of I against V. Repeat for the resistor.

Results

Your results can be recorded in a table.

Potential difference (V)	Current (I)
−2.0	−0.20
−1.5	−0.18
−1.0	−0.15
−0.5	−0.10
0	0
0.5	0.10
1.0	0.16
1.5	0.18
2.0	0.20

Method 2: Testing series and parallel circuits

Set up the circuits in both parallel and series arrangements and use these to verify the rules for the behaviour of current and potential difference in both types of circuits.

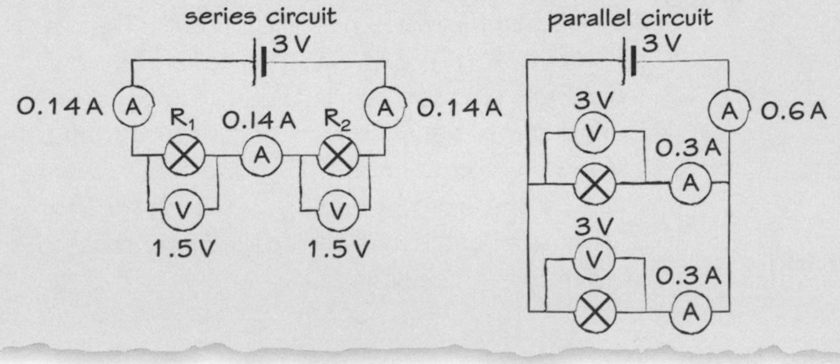

Conclusion

Current is the same at any point in a series circuit, but will split up at a junction in a parallel circuit.

The sum of the potential difference across components in any loop in a series or parallel circuit equals the potential difference of the cell.

There is more information relating to current, potential difference, resistance and series and parallel circuits on pages 80–85.

Be careful when using electrical circuits as electric current can produce heat and cause burns.

Data collected in an investigation are accurate if close to the true value. The data are precise if the repeat readings of a certain variable are close to one another. The best data is both accurate and precise.

In order to improve the accuracy of your results, place a switch in your circuit so that the current does not get too large. Too large a current leads to a greater value of the resistance being recorded, due to thermal energy causing extra vibrations of the ions in the metal lattice.

Now try this

1 Explain why a series or parallel arrangement will still lead to the same shape of I–V graph for both components. **(4 marks)**

2 Explain why the I–V graph for a filament lamp has the shape that it does. **(5 marks)**

The LDR and the thermistor

The resistance of **light-dependent resistors (LDRs)** and **thermistors** changes according to light conditions (LDR) or temperature (thermistor).

Light-dependent resistors

The resistance of a **light-dependent resistor (LDR)** is large in the dark. The resistance gets less if light shines on it. The brighter the light, the lower the resistance.

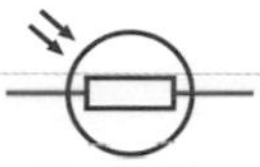

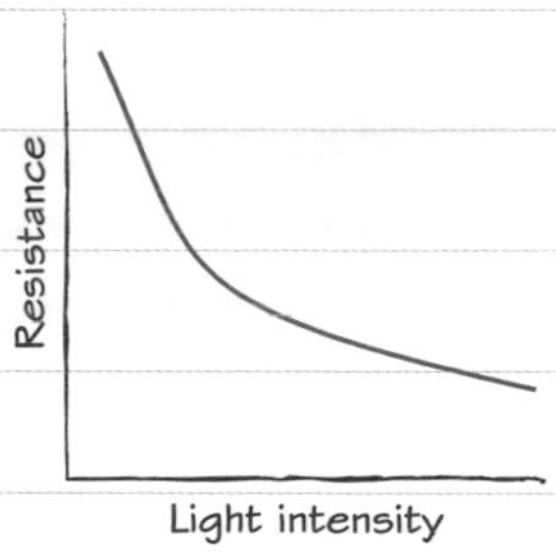

brightness **UP**, resistance **DOWN**

Thermistors

The resistance of a **thermistor** depends on its temperature. The higher the temperature, the lower the resistance.

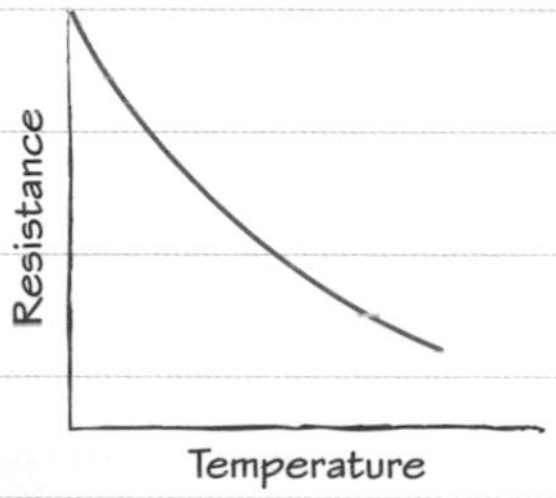

temperature **UP**, resistance **DOWN**

Worked example

A 6 V battery is connected in series with an LDR, a 300 Ω fixed resistor and an ammeter, as shown.

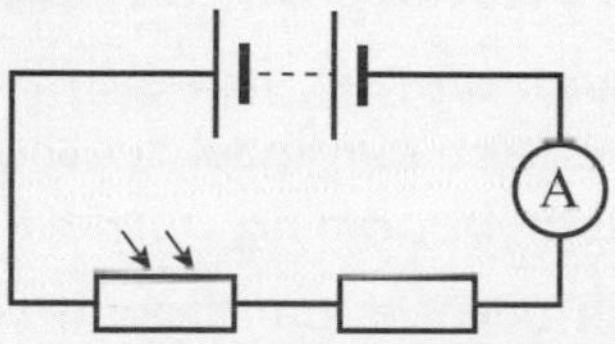

(a) Describe how the circuit shown can be used to explore the variation in resistance as the light levels change. **(2 marks)**

The amount of light being shone on the LDR can be changed by varying the intensity of the light it is exposed to from a torch or a lamp. The size of the current flowing in the circuit can then be recorded for each of these light levels.

(b) Describe the results you would expect. **(2 marks)**

The output current will be highest when the light level is greatest. As the brightness increases, the LDR's resistance decreases.

(c) How could a voltmeter be used in the circuit to show how the resistance changes with light levels? **(3 marks)**

Connect a voltmeter across the LDR. As the light level increases, the potential difference across the LDR will decrease since the resistance has decreased and a greater share of the 6 V will be across the fixed resistor.

Worked example

A 3 V battery is connected in series with a thermistor and an LDR, as shown.

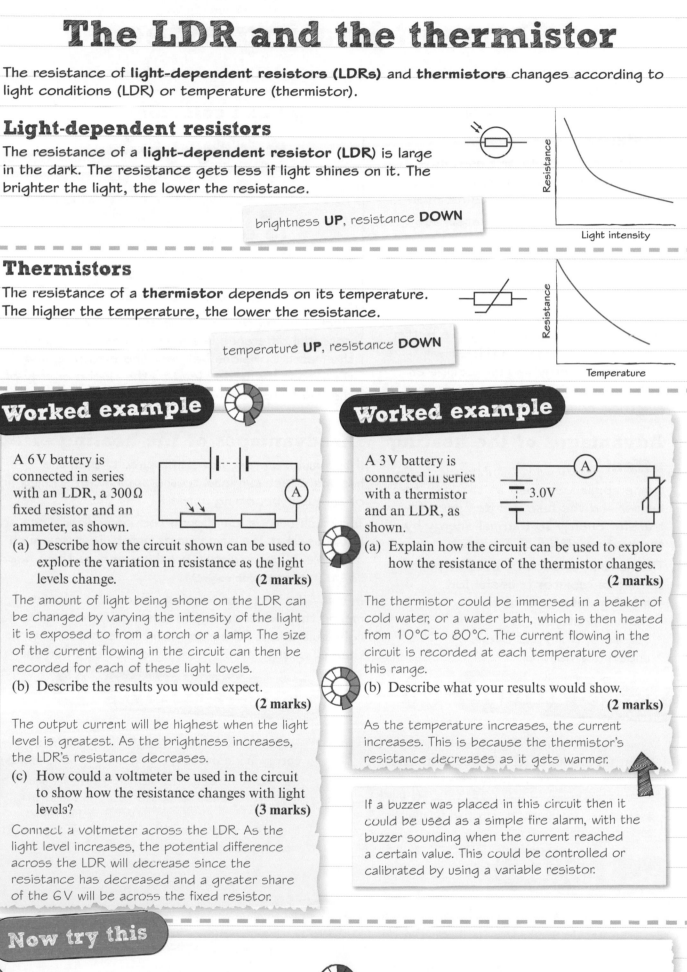

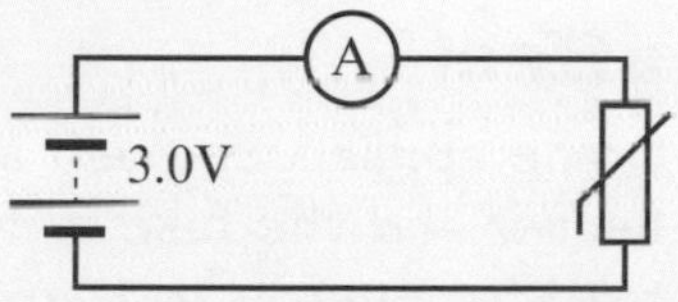

(a) Explain how the circuit can be used to explore how the resistance of the thermistor changes. **(2 marks)**

The thermistor could be immersed in a beaker of cold water, or a water bath, which is then heated from 10 °C to 80 °C. The current flowing in the circuit is recorded at each temperature over this range.

(b) Describe what your results would show. **(2 marks)**

As the temperature increases, the current increases. This is because the thermistor's resistance decreases as it gets warmer.

If a buzzer was placed in this circuit then it could be used as a simple fire alarm, with the buzzer sounding when the current reached a certain value. This could be controlled or calibrated by using a variable resistor.

Now try this

1 State what affects the resistance of (a) a thermistor and (b) an LDR. **(2 marks)**

2 Explain how an LDR and a thermistor could be used together in a circuit. **(4 marks)**

Current heating effect

When a **current** flows in a circuit it has a **heating** effect. This can have **advantages** and **disadvantages**.

Energy transfers in a resistor

When there is an electric current flowing through a resistor, energy is transferred which heats the resistor.

Electrical energy is **dissipated** as thermal energy in the surroundings when an electrical current does work against electrical resistance.

Unwanted thermal energy transfers can be reduced by using low-resistance wires. Wires that are better conductors, shorter or thicker will waste less energy as heat compared with longer, thinner wires of a poorer electrical conductor.

Energy and collisions

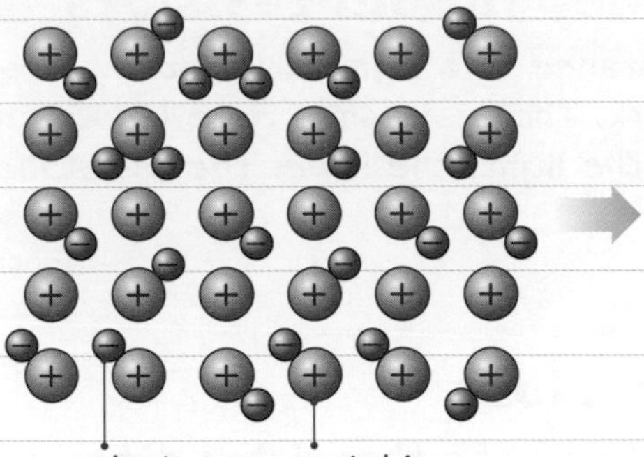

Electrons flow through a metal lattice when a potential difference is applied across the ends of the metal. Collisions between the electrons and the ions in the lattice lead to the kinetic energy of the electrons being dissipated as thermal energy.

Advantages of the heating effect

Some appliances such as the iron, the heater and the fuse are designed to transfer energy to thermal energy by the heating effect of a current.

The heating effect of a current passing through a resistor is useful for:

- heating water in a kettle
- radiant heaters
- toasters, grills and ovens
- underfloor heating.

Disadvantages of the heating effect

If too much current flows in a circuit, then the heating effect can lead to the appliance catching fire or the user becoming burned.

Too much current can flow if too many appliances are being used at the same time. Safety features such as correct fuses, circuit breakers and earthing need to be in place. See page 91.

In many devices, the heating effect of a current means that some energy is wasted. For example, a light bulb and a TV screen transfer some energy to thermal energy by the heating effect of a current.

Worked example

Give three examples of appliances in the home where the heating effect of the current causes energy to be wasted as thermal energy. **(1 mark)**

a computer, computer monitor and a mobile phone screen

Worked example

Explain how energy is transferred to thermal energy in a resistor. **(3 marks)**

Collisions between the moving electrons and the fixed metal ions in a lattice result in the kinetic energy of the electrons being transferred to thermal energy. The thermal energy is dissipated into the surroundings.

Now try this

1 Name three household devices that usefully transfer electrical energy to thermal energy. **(3 marks)**

2 Explain why it might be dangerous to have many electrical appliances being used at the same time. **(3 marks)**

3 Explain why it is better to use metals with a lower resistance in radiant heaters instead of metals with a higher resistance. **(3 marks)**

Energy and power

Electrical energy is **transferred** in circuits to do **work.** The amount of energy transferred depends on the **current**, **time** and **potential difference**.

Calculating energy

The total energy transferred by a device depends on the current in it, the potential difference across it and how long it is switched on.

Energy transferred (J) = current (A) × potential difference (V) × time (s)

$E = I \times V \times t$

> **Maths skills** Cover up the quantity you want to find with your finger. The position of the other quantities tells you the formula.
>
> $V = E \div (I \times t)$
>
> $I = E \div (V \times t)$
>
> $t = E \div (V \times I)$

Calculating power

Power is the energy transferred per second or the rate at which energy is transferred. It is measured in **watts**. For electrical devices, power depends on the current in the device and the potential difference across it, or the resistance of the device. It also depends on the total energy transferred and the time taken to transfer the energy.

> **Maths skills** The equation triangles for power can be used in the same way as for energy calculations.

There are three equations you need to know that can be used to calculate power:

1. electrical power (W) = current (A) × potential difference (V)
   $P = I \times V$
   LEARN IT! IT'S NOT ON THE EQUATIONS LIST

2. power (W) = energy transferred (J) ÷ time taken (s)
   $P = E \div t$
   LEARN IT! IT'S NOT ON THE EQUATIONS LIST

3. electrical power (W) = current squared (A^2) × resistance (Ω)
   $P = I^2 \times R$
   LEARN IT! IT'S NOT ON THE EQUATIONS LIST

Worked example

(a) A current of 8 A flows for 4 minutes in a kettle that is connected to a potential difference of 230 V. Calculate how much energy is transferred. **(4 marks)**

$E = V \times I \times t$

$E = 230\,V \times 8\,A \times (4 \times 60)\,s = 441\,600\,J$

(b) A 2.2 kW hairdryer is used for 5 minutes. Calculate how much energy is transferred. **(4 marks)**

$E = P \times t$

$= 2200\,W \times (5 \times 60)\,s = 660\,000\,J$

Worked example

(a) Calculate the power of a microwave that transfers 30 000 J of energy in 35 s. **(4 marks)**

$P = E \div t$

$= 30\,000\,J \div 35\,s = 857\,W$

(b) Calculate the current that is taken by a device with a power rating of 3400 W when its resistance is 120 Ω. **(4 marks)**

$I = \sqrt{(P \div R)} = \sqrt{(3400\,W \div 120\,\Omega)} = 5.3\,A$

Power can be measured in watts (W) or in joules per second (J/s). A power of 1 W means that 1 J of energy is being transferred each second.

Now try this

1 Calculate the current that flows in a 3000 W oven when connected to the 230 V mains supply. **(3 marks)**

2 A device transfers 100 000 J over 3 minutes. The potential difference across the device is 230 V. Calculate the resistance of the device. **(4 marks)**

A.c. and d.c. circuits

Circuits can be operated using **alternating current (a.c.)** or **direct current (d.c.)**.

Direct current

An **electric current** in a wire is a flow of electrons. The current supplied by **cells** and **batteries** is **direct current (d.c.)**. In a direct current the electrons all flow in the same direction.

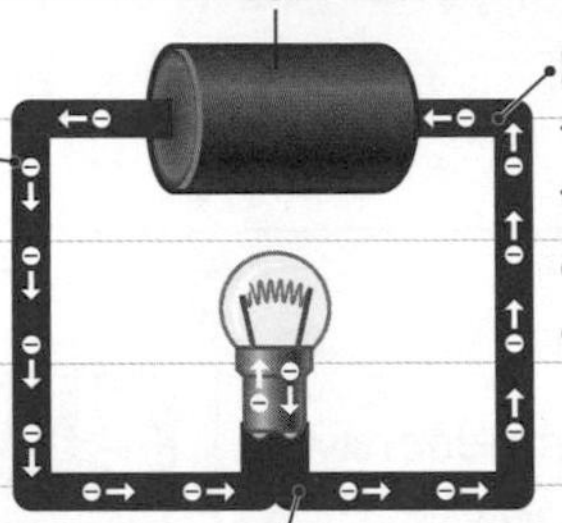

cell
Electrons are pushed out of one end of the cell.
Electrons flow round to the other end of the cell.

There must be a complete circuit for the electrons to flow.

The oscilloscope trace for d.c. from a battery is a horizontal line. The potential difference can be read off the vertical scale.

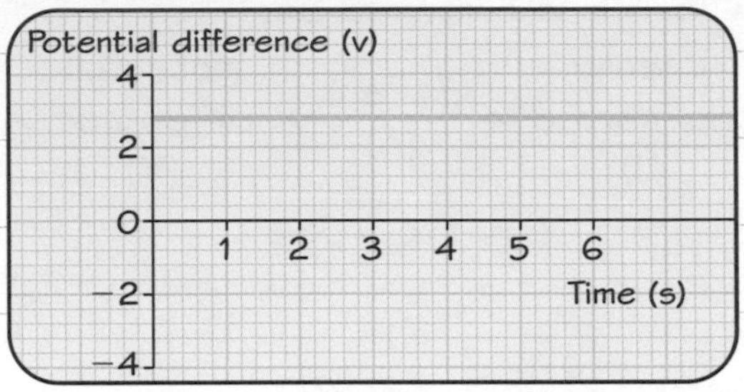

The potential difference above has a constant value of 2.8 V.

Alternating current

Mains electricity supplied to homes and businesses is an **alternating current (a.c.)**. Alternating current is an electric current that changes direction regularly, and its potential difference is constantly changing.

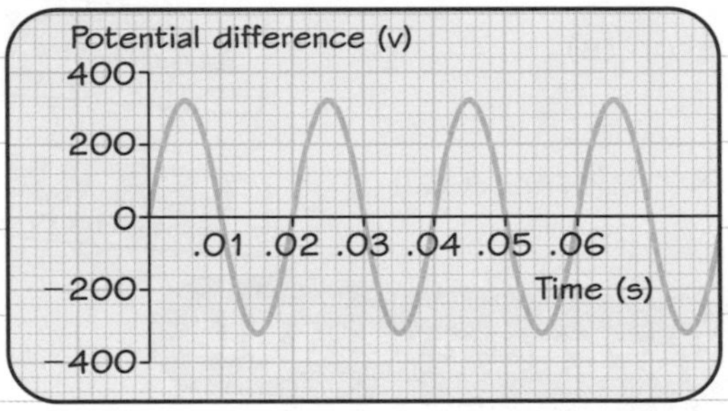

For an a.c. supply, the movement of charge is constantly changing direction. The mains supply has an average working value of 230 V and a frequency of 50 Hz. This means that the electric current and voltage change direction 100 times every second.

D.c. and a.c. supply in the home

Both d.c. and a.c. can be used in the home. D.c. is supplied in the form of cells and batteries. A.c. is the mains supply. Both can be used by devices to transfer energy to motors and heating devices.

Different electrical appliances have different power ratings. The power rating tells you how much energy is transferred by the appliance each second.

Appliance	Power rating
kettle	2200 W
hairdryer	1500 W
microwave	850 W
electric oven	3000 W
electric shaver	15 W

Worked example

Compare the power rating of a hairdryer and an electric shaver, and the changes in stored energy when they are in use. **(4 marks)**

The hairdryer has a power rating of 1500 W, which means that it transfers 1500 J per second to stores of kinetic energy and thermal energy. The electric shaver has a power rating of one-hundredth of the hair dryer. It transfers only 15 J per second, mainly to a kinetic energy store.

Now try this

1 Give three examples of devices that use
(a) d.c. **(3 marks)**
(b) a.c. **(3 marks)**

2 Explain how a.c. and d.c. are
(a) similar **(1 mark)**
(b) different. **(2 marks)**

3 Explain how and why alternating current is generated in power stations instead of direct current. **(3 marks)**

Mains electricity and the plug

Electrical energy enters UK homes as **mains electricity** at **230 V a.c.**. A mains plug has three wires, which have different roles in the operation of a.c. in the home.

The 13 A plug

The earth (yellow and green) wire does not form part of the circuit but acts as a safety feature along with the fuse.

The neutral (blue) wire completes the circuit with the appliance. It is at a potential difference of 0 V with respect to the earth wire.

The live, neutral and earth wires are made from copper, but covered in colour-coded insulating plastic so that the consumer is not exposed to a dangerous voltage.

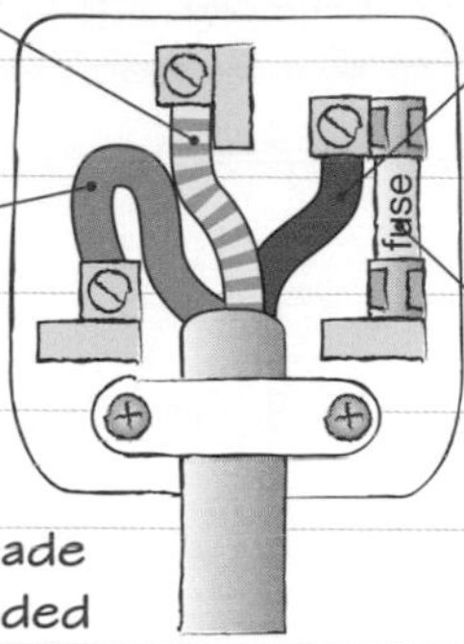

The live (brown) wire carries the supply to the appliance. It is at a potential difference of 230 V with respect to the neutral and earth wires.

The fuse is connected to the live wire and it contains a wire that will melt if the current gets too high. If the fuse blows, the device will be at 0 V and not at 230 V. A switch should also be connected in the live wire for the same reason.

Earthing and fuses

Earthing, circuit breakers and fuses are safety features that are used to ensure that the user does not get electrocuted if a fault occurs.

1 The live wire inside the appliance may come loose and touch a metal part of the device's casing.

2 The large current heats and melts the wire in the fuse, making a break in the circuit.

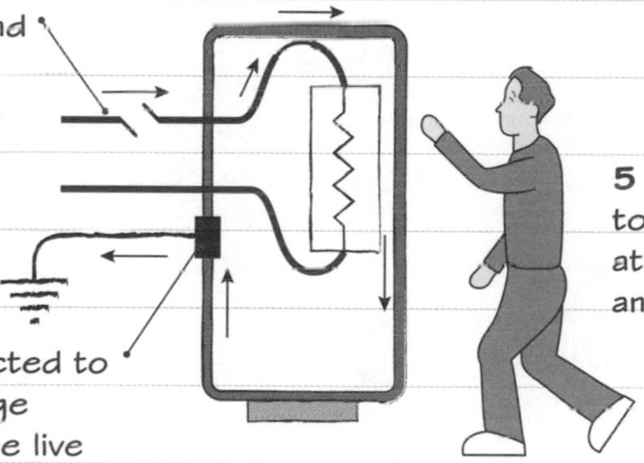

5 By having an earth wire connected to the metal casing, the user is not at risk if the live wire comes loose and touches anything metallic.

3 The earth wire is connected to the metal casing and a large current flows in through the live wire and out through the earth wire.

4 The circuit is no longer complete, so there is no chance of electric shock or fire.

Worked example

(a) State the potential differences of the live, neutral and earth wires in a plug. **(3 marks)**

live: 230 V; neutral and earth: both at 0 V

(b) Name three safety features used by appliances. **(3 marks)**

earth wire, fuse and circuit breaker

Circuit breakers are also used as safety features in addition to fuses and earth wires. When circuit breakers detect that a current is too high, they use a magnetic field to open a switch and isolate the appliance, making it safe. They are easily reset by pressing a switch if they trip.

Now try this

1 State the colours of the
(a) live wire **(1 mark)**
(b) neutral wire **(1 mark)**
(c) earth wire. **(1 mark)**

2 Give one advantage of a fuse and one advantage of a circuit breaker. **(2 marks)**

3 Explain how an earth wire and fuse protect the user when the live wire comes loose. **(4 marks)**

Extended response – Electricity and circuits

There will be one or more 6 mark questions on your exam paper. For these questions, you will need to think scientifically and structure your answer logically, showing how the points you make are related to each other. You can revise the topics for this question, which is about **electricity and circuits**, on pages 45 and 79–91.

Worked example

Two identical bulbs can be connected to a 6 V cell in series or parallel as shown in the diagram.
Compare the brightness of the bulbs in the two circuits when the switches are closed.
Your answer should refer to the current, potential difference, energy stores and rates of energy transfer in both circuits.

(6 marks)

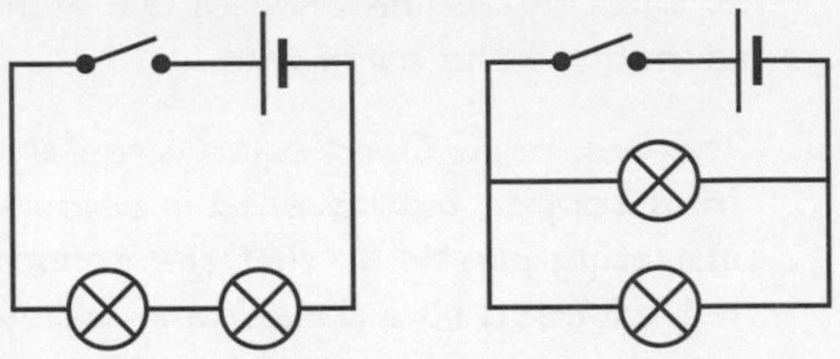

The brightness of the bulbs in the series circuit will be much less than the brightness of the bulbs in the parallel circuit.

The electrical energy that is transferred to light is directly proportional to the potential difference across each bulb and the size of the current flowing through each bulb. The brightness of the bulbs will be greatest when the energy transferred per second is greatest.

In the series circuit, the potential difference across each bulb will be 3 V, whereas in the parallel circuit the potential difference across each bulb will be 6 V. Also, the current flowing through each of the bulbs in the parallel circuit will be greater than the current flowing through the bulbs in the series circuit. Since the current and the potential difference in the parallel bulbs are greater, more energy will be transferred per second and the bulb will be brighter.

The opening sentence simply states how the brightness of the lamps compares in each of the two circuits.

The answer then states the relationship between electrical energy, current and potential difference before relating brightness to the energy transferred per second, or power, of the bulbs.

The comparison of the bulbs' brightness is related to the relative sizes of any current, potential difference and energy transferred per second.

The rate at which energy is transferred from one form to another is called power. When dealing with electrical circuits, the power is calculated using the equation $P = IV$. You can read more about this on page 89.

Now try this

The UK mains electricity supply operates at 230 V a.c. Explain how safety features help to keep users safe. Your answer should refer to the structure and function of a plug. **(6 marks)**

Static electricity

Static electricity occurs when electric charges are transferred **onto**, or **off**, the surface of an **insulator**. This causes the insulator to gain a **positive** or a **negative** charge.

Negatively charged insulators

Insulators become **negatively charged** when **electrons** move **onto** the **insulator** from the cloth due to **friction**.

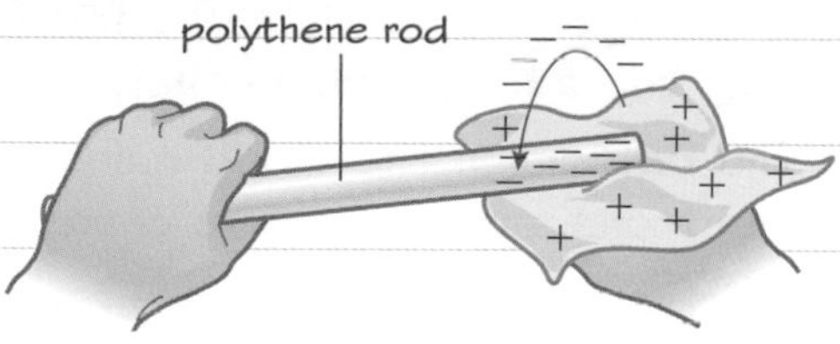

When **polythene** is rubbed with a cloth, electrons are transferred by **friction** from the **cloth** to the **rod**. Since electrons are **negatively** charged, the rod becomes **negatively charged** and the cloth becomes **positively charged**. These charges are **equal** and **opposite**.

Negative rod, positive cloth

It is **always** the **electron** that is transferred when insulators become charged by friction.

Positively charged insulators

Insulators become **positively** charged when electrons move **off** the **insulator** and onto the cloth due to the force of **friction**.

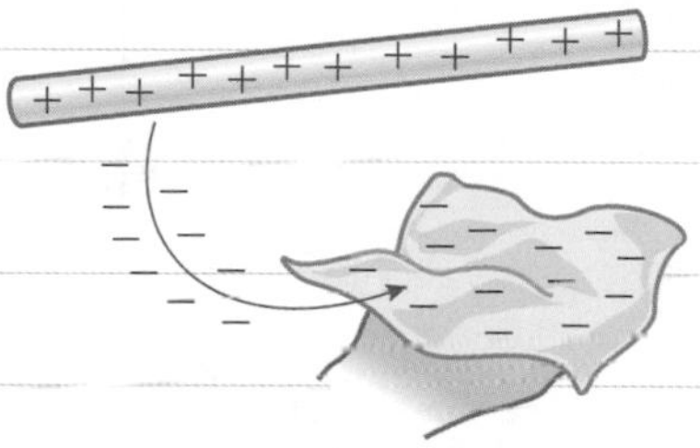

When **acetate** is rubbed with a cloth, electrons are transferred by **friction** from the **rod** to the **cloth**. Again, since electrons are negatively charged, the **cloth** becomes **negatively charged** and the **acetate rod** becomes **positively charged**. Again, these charges will be **equal** and **opposite**.

Positive rod, negative cloth

Worked example

Describe what happens when two electrically charged objects are brought close together. **(3 marks)**

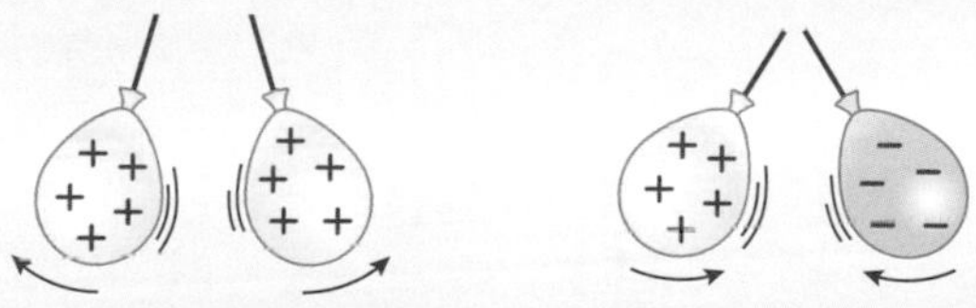

The charged objects exert a force on each other. If the two objects have the same type of electrical charge they repel each other. If they have different types of electrical charge they attract each other.

Now try this

1 What charge will an insulator have if it
(a) gains electrons **(1 mark)**
(b) loses electrons? **(1 mark)**

2 Which objects will attract each other?
☐ **A** two negatively charged rods
☐ **B** a positively charged duster and a negatively charged rod
☐ **C** two positively charged objects
☐ **D** a neutral rod and a neutral duster **(1 mark)**

3 Explain why insulators never become charged by gaining or losing protons. **(3 marks)**

4 Polythene gains electrons when rubbed. Explain how you could use a polythene rod to tell if an insulator were positively charged. **(3 marks)**

Electrostatic phenomena

Electrostatic behaviour can be seen in action every day.

Electric shocks

Electrons can be transferred onto your clothes by friction. They stay on your clothes because the materials from which clothes are made are insulators. However, if you then touch a metal handle, the electrons can conduct to earth and you will receive an electric shock. This an example of earthing.

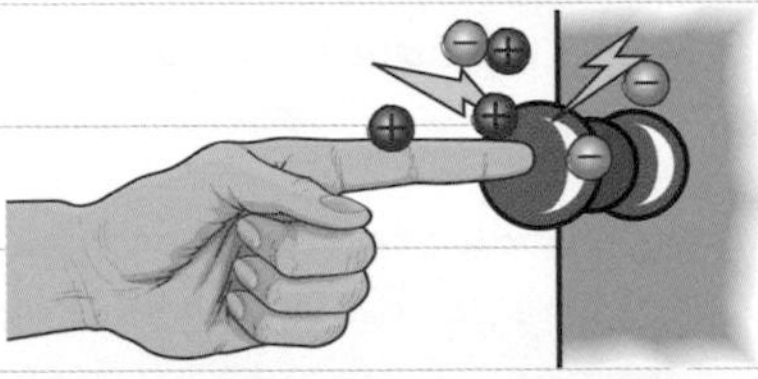

Lightning strikes

Lightning occurs because of electrostatic induction:

1. Ice particles in clouds gain electrons from other ice particles by friction as they rub against each other.
2. The bottom of the cloud gains electrons and a negative charge.
3. Electrons in the ground are repelled by the bottom of the cloud.
4. Lightning jumps to earth as a spark of electrons.

Electrostatic induction

A **charged object** (such as a plastic comb) can attract **uncharged objects** (such as small pieces of paper). This happens because the comb **induces** a charge in the pieces of paper.

Electrostatic induction involves an opposite charge being induced on the surface of a neutral insulator by another charged insulator. In both cases, the force of attraction is seen.

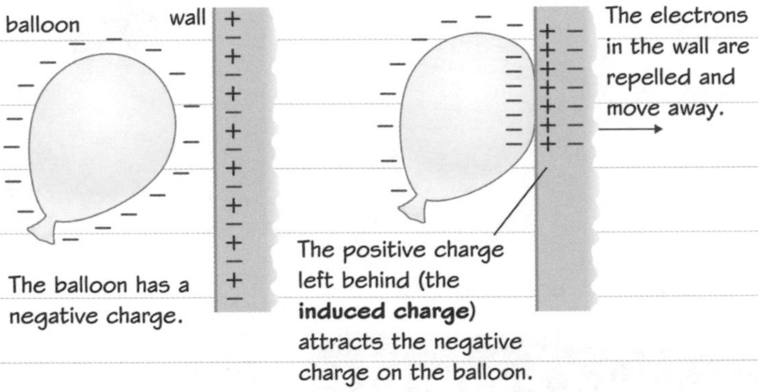

balloon
wall
The balloon has a negative charge.
The electrons in the wall are repelled and move away.
The positive charge left behind (the **induced charge**) attracts the negative charge on the balloon.

Worked example

Explain how small pieces of paper are attracted by electrostatic induction by a comb when the comb is rubbed with a cloth. **(4 marks)**

Friction occurs between the cloth and the comb.

Electrons are transferred between these insulators so that one becomes positive and the other one negative.

When the comb is placed near the paper, the opposite charge is induced on the paper.

The opposite charges attract, so the paper is attracted to the comb and is lifted upwards, overcoming the gravitational force acting on it.

Electrostatics questions often require friction, electrons, insulators, induction and attraction/repulsion to be mentioned.

This only happens with small pieces of paper where the force of electrostatic attraction is stronger than their weight.

Now try this

1 Explain why a comb can be statically charged if it is plastic but not if it is metal. **(2 marks)**

2 Explain why you can get an electric shock if you touch a door handle. **(5 marks)**

3 Lightning strikes can occur between clouds. Explain how this happens. **(4 marks)**

Electrostatics uses and dangers

Static electricity can be very useful but also very dangerous.

Uses of electrostatics

Static electricity is useful in **paint spraying** and **insecticide** sprayers.

Insecticide sprayers

Nozzle of sprayer is connected to electricity supply → Droplets all get the **same kind** of static charge → Droplets **repel** each other so the spray spreads out evenly

Paint sprayers

Nozzle of sprayer is connected to electricity supply → Droplets all get the **same kind** of static charge so they spread out evenly → The object being painted is given the **opposite** charge to the paint → The paint is attracted to the object being painted and less paint is wasted

Dangers of electrostatics

When tankers, cars or aircraft move, friction occurs between the body of the vehicle and the air. This results in electrons being transferred and the vehicle becoming charged.

Friction between fuel and the pipe it is flowing through can also cause the transfer of electrons so that the fuel and pipe become charged.

Earthing

If the charge is not removed from an aeroplane or tanker, then a spark may occur when the nozzle of the fuel tanker touches the aeroplane. The spark is an uncontrolled discharge of electrons, and it can ignite fuel vapours and cause an explosion.

Earthing is used to ensure safe movement of electrons off the surface to avoid a dangerous build-up of charge.

An earthing or bonding wire is connected between the aeroplane or fuel tanker and ground before refuelling starts.

Worked example

Electrostatics is used in the spray-painting of cars. The car frame is charged. Suggest how the process works. **(5 marks)**

The paint droplets become charged due to friction and will repel as they all have the same type of charge. The droplets spread out and cover all of the car frame, which has been given the opposite charge. This means that there will be an even coating of paint, and less paint will be wasted.

Worked example

Explain how static electricity can be dangerous when refuelling an aeroplane and how the problem is solved. **(4 marks)**

Aeroplanes become charged due to friction when they fly, as electrons are transferred from the air to the outer casing of the plane. When they land, the charge needs to be removed by taking electrons to the ground with an earthing wire. If this is not done, the charge may cause a spark, which can ignite the fuel vapour, resulting in an explosion.

Now try this

1 Explain why insecticide spray droplets spread out when they are charged. **(2 marks)**
2 Explain the main stages in the use of an electrostatic insecticide spray. **(4 marks)**
3 Explain why charge can be dangerous when refuelling a plane. **(3 marks)**

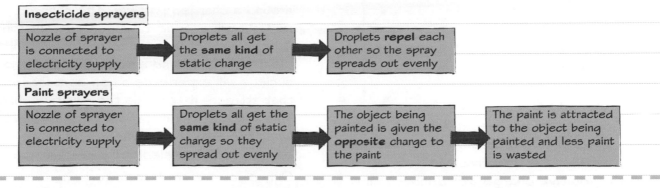

Electric fields

An **electric field** is a region in space where a **charged particle** may experience a **force**.

Electric field from a point charge

The **electric field** from a **positive point charge** acts **radially outwards**.

The **electric field** from a **negative point charge** acts **radially inwards**.

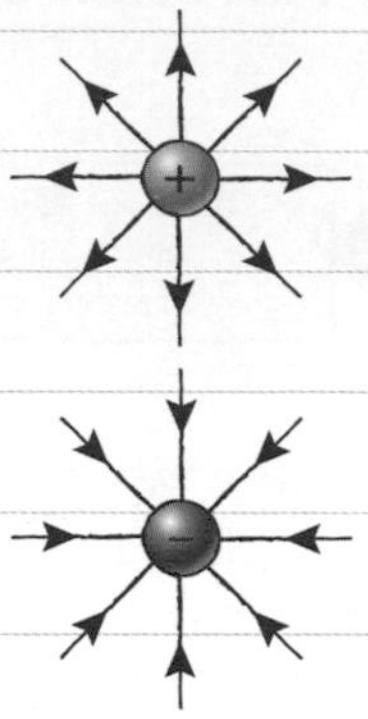

Electric field strength is a **vector** quantity because it has both a **size** and a **direction**. The arrows show the direction that a **positive charge** would move in if it were placed in the field.

A positive charge in the field around a negative charge would move towards the negative charge, since positive and negative charges attract.

An electric field is created in the region around a charged particle. The charged particle will experience a non-contact force. A negatively charged electron placed in the electric field around a positively charged nucleus will experience a non-contact force that attracts it towards the positive charge. This explains why materials can gain electrons when they are rubbed, and explains why we encounter static electricity in nature.

Only electrons are transferred when an insulator becomes charged. If a material gains electrons it becomes negatively charged; if it loses electrons it becomes positively charged.

Showing the strength of the field

The strength of an electric field is shown by the concentration of the field lines. The more concentrated the lines, the stronger the electric field.

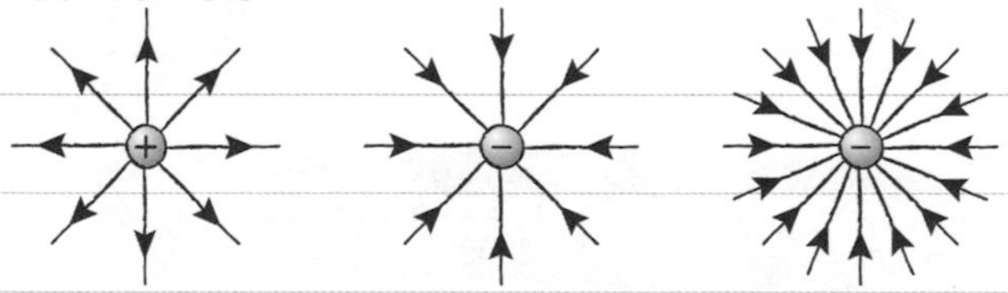

weak positive point charge | weak negative point charge | strong negative point charge

The electric field between parallel plates

The electric field lines between oppositely charged parallel plates are parallel lines.

The more concentrated the field lines, the stronger the electric field.

The field lines are curved at the ends of the plates.

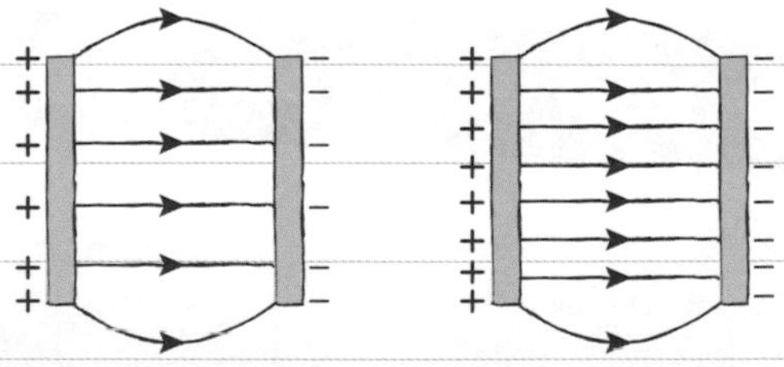

Worked example

Explain what will happen to a proton that is placed in the electric field of a point positive charge. **(3 marks)**

A proton is a positively charged particle, so it will be repelled by the positive charge and move outwards. Since the proton will experience a resultant force, it will accelerate.

Worked example

Describe the path taken by an electron that is placed between oppositely charged parallel plates. **(3 marks)**

The electron is negatively charged and will be repelled by the negative plate and attracted by the positive plate. This will cause it to accelerate towards the positive plate.

Now try this

1 Explain how the motion of a charged particle changes in an electric field when:

(a) its mass changes **(2 marks)**

(b) it is given the opposite charge. **(2 marks)**

2 Explain how electric fields provide an explanation of static electricity. **(4 marks)**

Extended response – Static electricity

There will be one or more 6 mark questions on your exam paper. For these questions, you will need to think scientifically and structure your answer logically, showing how the points you make are related to each other. You can revise the topics for this question, which is about **static electricity**, on pages 93–96.

Worked example

An electric field is a region in space where an electric charge experiences a force.

There are two common shapes for an electric field: a radial field around a point charge and the uniform parallel field between metal plates, as shown in the diagrams.

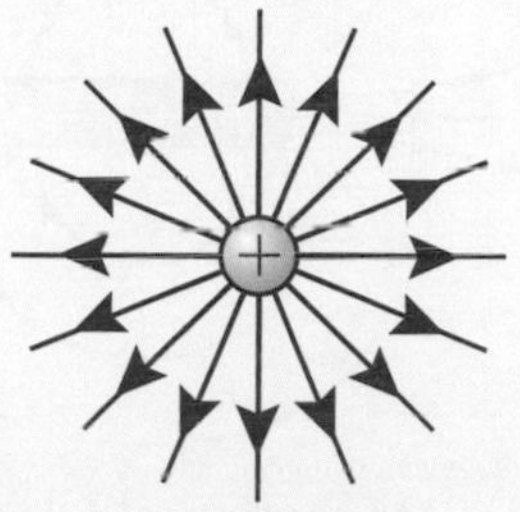

radial field

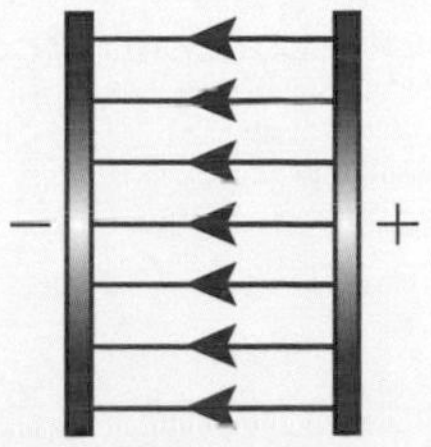

uniform field

Explain the motion of a positively charged proton in each of these fields.

Your answer should refer to the size and direction of the forces acting on the proton in each of the electric fields shown. **(6 marks)**

In the case of the radial field, the proton will move away from the positive point charge because like charges repel. Similarly, the proton will move away from the positive plate and towards the negative plate when placed in the region of the uniform field. It will be repelled by the positive plate and attracted by the negative plate.

In both cases, the direction of the proton's motion is stated, with a simple explanation given based on the nature of repulsion and attraction of charged particles.

The force acting on the proton in the radial field will be radially outwards from the point charge and the size of the force will decrease in magnitude the further the proton moves away from the point charge. This is shown by the fact that the field lines diverge (get further apart) as the distance from the point charge increases. For the uniform field, the force is constant throughout and acts from the positive plate to the negative plate.

The magnitude of the force on the proton is being related to size of the field as well as the shape and the direction of the field lines.

The acceleration of the proton in the radial field will decrease as it moves further away from the point charge. For the uniform field, the force will be constant throughout as indicated by the parallel lines, so the particle will have a constant acceleration between the plates.

The relationship between force and acceleration is given by Newton's second law and the equation $F = ma$. If the force acting on the body gets smaller then the acceleration will also get smaller.

You can find out more about the relationship between force and acceleration on page 8.

Now try this

A metal rod, a Perspex rod and a polythene rod can all be rubbed with a cloth.

Explain why certain materials, when rubbed, will attract pieces of tissue paper whereas others will not.

Your answer should refer to the nature of insulators and conductors and the movement of particles when materials become statically charged. **(6 marks)**

Magnets and magnetic fields

Certain materials can be magnetised to become **permanent** or **temporary magnets**.

Magnets and magnetic fields

Like magnetic poles **repel**. **Unlike** magnetic poles **attract**.

The magnetic field is strongest at the poles. The field lines are shown to be closer together and more concentrated.

bar magnet

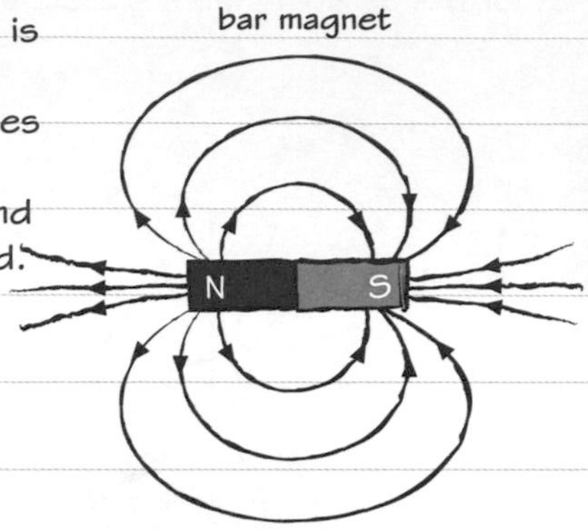

Magnetic field lines are always from N to S.

uniform field

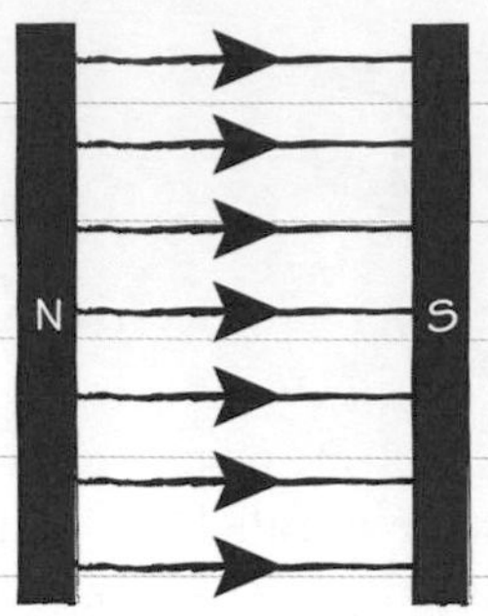

The field is constant. This is shown by parallel, equally spaced field lines.

Permanent and induced magnets

Magnetic materials include **cobalt, steel, iron, nickel** and magnadur.

A **permanent magnet** has poles which are N and S all of the time. They are made from steel or magnadur.

A **temporary magnet** can be magnetised by bringing a permanent magnet near to it. When the permanent magnet is removed, the temporary magnet loses its magnetism.

Uses of magnets

Permanent	Temporary
fridge magnets	electromagnets
compasses	circuit breakers
motors and generators	electric bells
loudspeakers	magnetic relays
door closers on fridges	

Plotting compasses

A plotting compass is used to plot the shape and direction of magnetic lines of force. You can do this by laying a bar magnet on a piece of paper. Put the plotting compass by one pole of the magnet. Draw a dot by the needle away from the magnet. Move the plotting compass so that the other point is by the dot you have just drawn. Continue doing this until you have mapped all the way around the magnet.

The Earth's magnetic field has the same pattern as that of a bar magnet and can be plotted using a plotting compass. The behaviour of compasses is evidence that the Earth has a magnetic field.

Worked example

What will the plotting compass needle look like when it is placed at point 1? **(1 mark)**

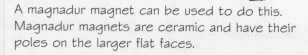

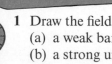

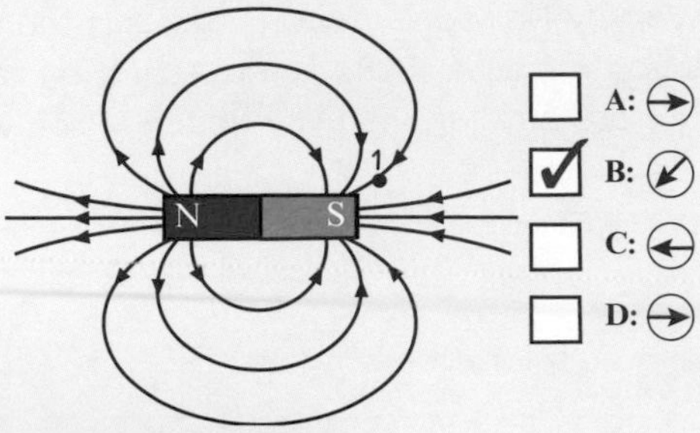

Explain how you can tell if a magnetic material is a permanent or a temporary magnet. **(2 marks)**

A temporary magnetic material will always be attracted by a magnet, but a permanent magnet can be both attracted and repelled.

A magnadur magnet can be used to do this. Magnadur magnets are ceramic and have their poles on the larger flat faces.

Now try this

1 Draw the field lines for
(a) a weak bar magnet **(2 marks)**
(b) a strong uniform field. **(2 marks)**

2 Explain why unmagnetised iron is always attracted by a permanent magnet. **(3 marks)**

3 Suggest why motors use permanent magnets. **(2 marks)**

Current and magnetism

An **electric current** will create a **magnetic field.** The **shape, direction** and **strength** of the field depend on a number of factors.

The magnetic field around a long straight conductor

The shape of the magnetic field around a long straight conductor can be thought of as a series of concentric circles.

Be careful! Do not confuse the direction of the current with electron flow. You need to use the direction of conventional current which is taken to be from + to –.

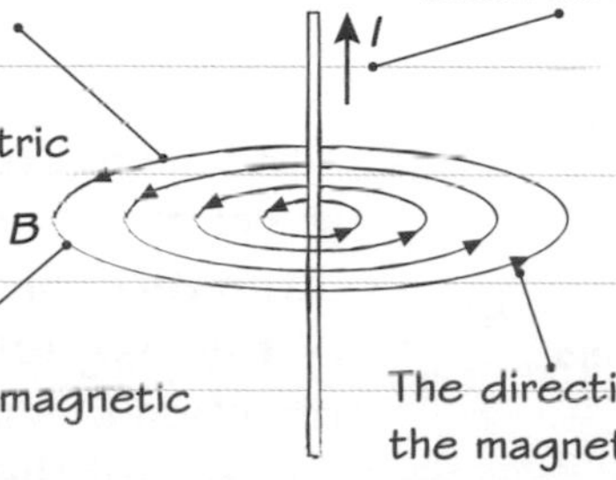

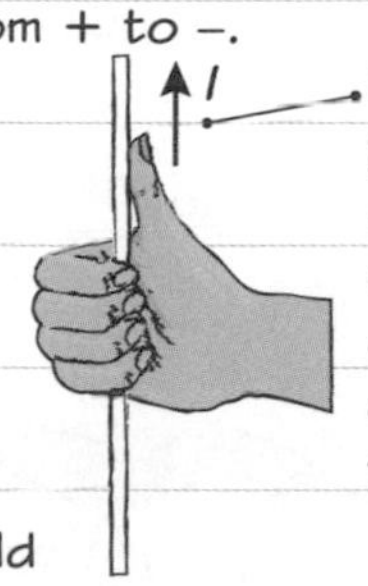

If you point the thumb of your right hand in the direction of the flow of conventional current, the direction that your fingers curl will be the direction of the magnetic field.

The strength of the magnetic field depends on:
- the size of the current in the wire: it is directly proportional to the current.
- the distance from the wire: it is inversely proportional to the distance from the wire.

The direction of the magnetic field depends on the direction of the electric current in the conductor.

The solenoid

The magnetic field lines of the individual coils in a solenoid add up to give a very strong, uniform field along the centre of the solenoid. However, the field lines cancel to give a weaker field outside the solenoid.

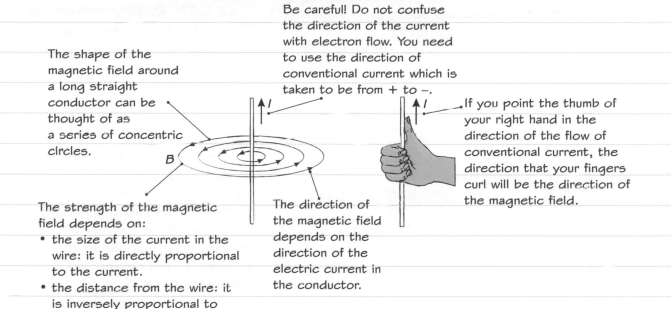

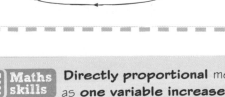

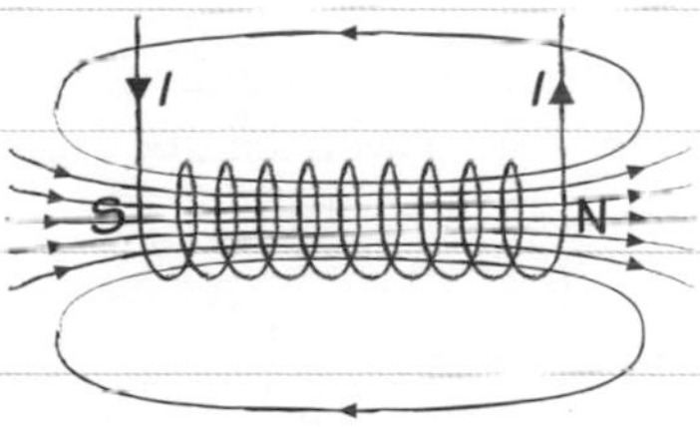

Worked example

Explain how the strength of a magnetic field due to a current in a long straight wire changes when:

(a) the current increases **(1 mark)**

It will increase.

(b) the distance increases. **(1 mark)**

It will decrease.

Maths skills **Directly proportional** means that as **one variable increases** the other **variable increases** at **the same rate.** If the current doubles, the magnetic field strength will also double.

Maths skills **Inversely proportional** means that as **one variable increases** the other **decreases**. If the distance doubles, the field strength halves.

Now try this

1 Draw the magnetic field around a wire when the current is flowing downwards. **(2 marks)**

2 The current flowing in a long straight wire and the distance from the wire are both doubled. Explain what effect this will have on the magnetic field strength. **(3 marks)**

3 Suggest how the magnetic field strength inside a solenoid could be increased. **(3 marks)**

Current, magnetism and force

A current-carrying conductor, placed near a magnet, will experience a force due to the interaction of their magnetic fields.

Fleming's left-hand rule and the motor effect

An electric current flowing through a wire has a magnetic field. If you put the wire into the field of another magnet the two fields affect each other and the wire experiences a force. This is called the **motor effect**.

The maximum force on the wire occurs when the current is at right angles to the lines of the magnetic field. We can work out the direction of the force using **Fleming's left-hand rule**.

This rule uses 'conventional current' which flows from the + to the − of a cell. This is the opposite direction to the flow of electrons.

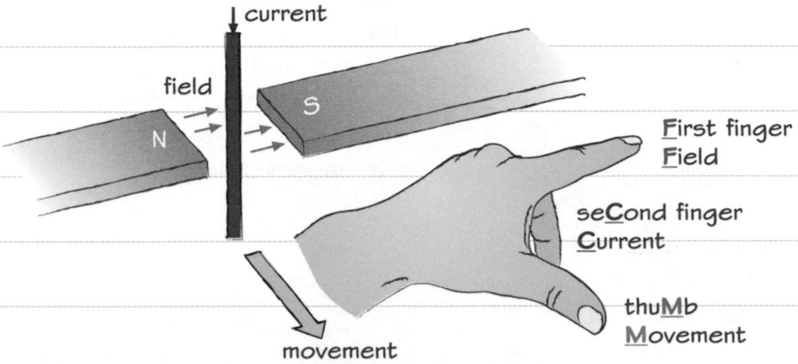

- There is **no force** if the **current** is **parallel** to the **field lines**.
- If the **direction** of either the **current** or the **magnetic field** is reversed the direction of the **force** is reversed.
- The size of the **force** can be **increased** by **increasing** the **strength** of the **magnetic field**, or **increasing** the **size** of the **current**.

Worked example

An electric motor uses the motor effect to make a motor go round. Use the diagram to help you answer the questions.

(a) Explain why the left-hand side of the coil moves upwards. **(2 marks)**

The left-hand side of the coil is pushed upwards by the motor effect. The magnetic field runs from left to right, the current is running out of the page and we can use Fleming's left-hand rule to work out that the direction of the force is upwards.

(b) Explain why the coil rotates. **(2 marks)**

The motor effect means that the left-hand part of the coil moves up and the right-hand side moves down. This makes the coil spin. The split ring swaps the connections to the battery over every half turn, so the coil continues to spin in the same direction.

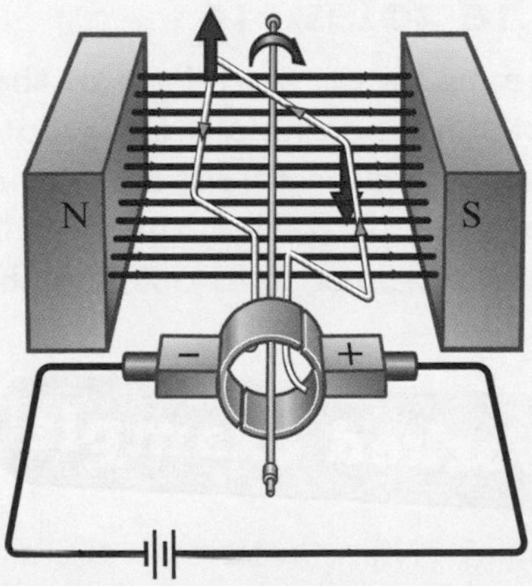

Force on a current-carrying wire

You can calculate the force on a wire using the equation:

force on a conductor at right angles to a magnetic field carrying a current (N) = magnetic flux density (T or N/A m) × current (A) × length (m)

$$F = B \times I \times L$$

Now try this

1 Explain why a current-carrying wire experiences a force when placed near a magnet. **(2 marks)**

2 Calculate the force acting on a wire when $B = 1.2\,\text{mT}$, $I = 3.6\,\text{A}$ and $L = 50\,\text{cm}$. **(3 marks)**

3 For an electric motor, explain how the (a) direction of rotation **(2 marks)** and (b) speed of rotation **(3 marks)** can be changed.

Extended response – Magnetism and the motor effect

There will be one or more 6 mark questions on your exam paper. For these questions, you will need to think scientifically and structure your answer logically, showing how the points you make are related to each other. You can revise the topics for this question, which is about **magnetism** and the **motor effect**, on pages 98 to 100.

Worked example

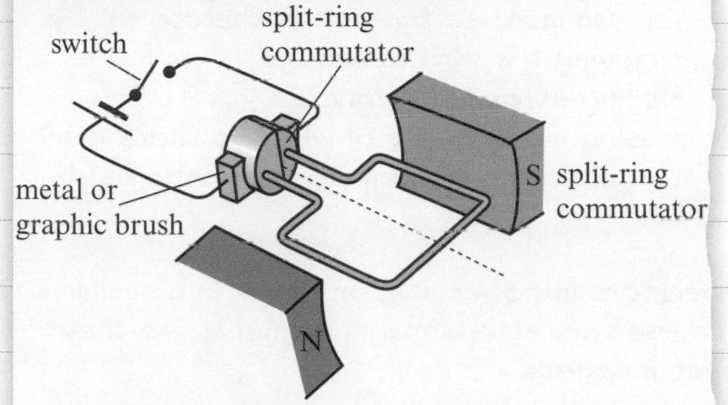

The diagram shows an electric motor.
Explain what will happen to the coil when the switch is closed.
Your answer should refer to the direction of the magnetic field, current and any forces acting on the coil of wire. **(6 marks)**

Conventional current flows from the positive terminal of a cell to the negative terminal. In this case, this means that the current will flow anticlockwise in the external circuit and in the coil between the two magnetic poles.

To start the answer, the direction of conventional current is stated. Do not consider the direction of the flow of electrons when answering questions relating to electric motors.

Magnetic field lines always come out of an N pole and go into an S pole. In the diagram, this will be a uniform field acting horizontally from the N pole on the left to the S pole on the right.

A short statement is made referring to the direction of the field and its uniform nature.

The coil will rotate in a clockwise direction. By applying Fleming's left-hand rule, there will be an upward force on the left arm of the coil and a downward force on the right arm of the coil. This will produce a couple, leading to clockwise rotation of the coil.

This part of the answer is the most detailed, since the explanation of the rotation of the coil is being linked to Fleming's left-hand rule and the behaviour of moments, couples and rotation.

Command word: Explain

When you are asked to **explain** something, it is not enough to just state or describe something. Your answer **must** contain some reasoning or justification of the points you make based on the appropriate physics. This needs to be communicated clearly using good spelling, punctuation and grammar.

Now try this

Magnetism can be used to attract a variety of materials of different masses. Explain how the strength of the magnetic field can be changed to make this possible. Your answer should refer to how the size of the magnetic field strength can be changed **with** and **without** the use of an **electric current**. **(6 marks)**

Electromagnetic induction

Magnets and **conductors** are used to generate alternating current and direct current.

Inducing a current in a wire

If you **move** part of a loop of wire in a **magnetic field**, an electric current will flow in the wire. This is called **electromagnetic induction**, and the current is an **induced current**.

You can get the same effect by keeping the wire still and moving the magnet.

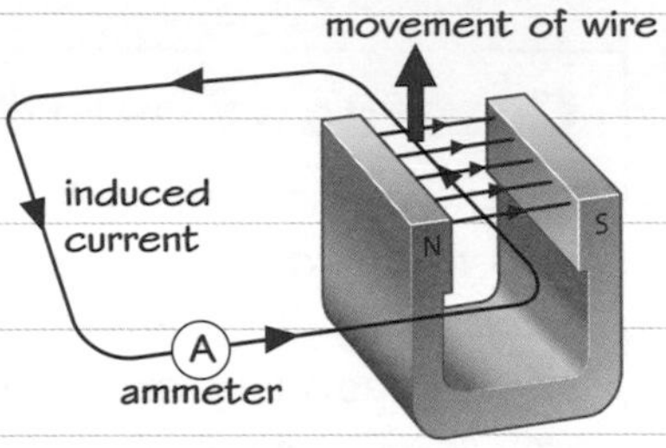
movement of wire
induced current
N
S
A
ammeter

Factors that affect the induced voltage

You can change the direction of the current by:

- changing the **direction of motion** of the wire
- changing the **direction** of the **magnetic field**.

You can increase the size of the current by:

- moving the wire **faster**
- using **stronger** magnets
- using **more loops** of wire, so there is more wire moving through the magnetic field.

Generators

This diagram shows a simple **generator**. The generators in power stations work in a similar way to the one in the diagram. However, they need to use very strong magnetic fields, so they usually use **electromagnets** instead of **permanent magnets**.

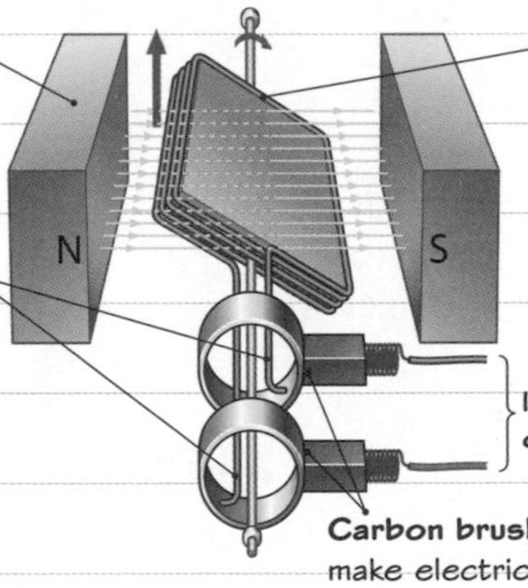
Permanent magnets produce a magnetic field. The stronger the magnetic field, the greater the current.
The ends of the coil are connected to **slip rings**. These allow the coil to spin without twisting the wires to the rest of the circuit.
N
S
A coil is wound on an **iron core**. The many turns of wire and the iron core increase the size of the current.
Induced **alternating current (a.c.)**
Carbon brushes press on the slip rings to make electrical contact with the rest of the circuit.

Worked example

(a) Explain how an alternator produces an a.c. output. **(4 marks)**

As the coil rotates, it cuts through the lines of magnetic field and this induces a current in the wire. The slip rings form an electrical contact with the carbon brush to allow current to flow in an external circuit. Every half turn the position of the coil leads to a reversal in the direction of the induced current.

A dynamo works in the same way as the alternator. Every half turn, contact with the carbon brushes is broken due to the gap in the ring. As it continues to rotate, the induced current is always in the same direction because the direction of the cutting of the field is always upwards on the left and downwards on the right.

Now try this

1 What is needed for electromagnetic induction to take place? **(3 marks)**

2 Explain how the alternator in a power station is different from one in a school laboratory. **(3 marks)**

Microphones and loudspeakers

Microphones and **loudspeakers** both depend on **sound energy** and **electrical energy**, but in different ways.

The microphone

A **microphone** transfers **sound waves** into an **electrical signal**.

As the diaphragm moves in and out, a varying potential difference is induced across the ends of the wire, producing a varying electric current.

The sound wave is made up of areas of high pressure and low pressure.

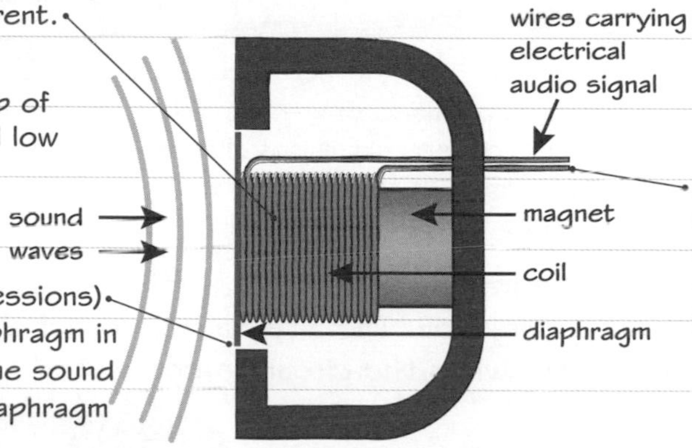

The high pressure areas (compressions) of the sound wave push the diaphragm in and the low pressure areas of the sound wave (rarefactions) cause the diaphragm to move outwards.

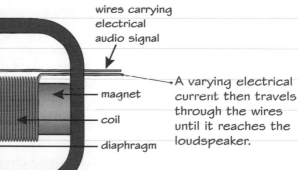

A varying electrical current then travels through the wires until it reaches the loudspeaker.

The loudspeaker

A **loudspeaker** converts an **electrical signal** into **sound waves**.

The coil of wire is wrapped around an iron core and placed between the poles of a permanent magnet.

The varying potential difference leads to a varying current in the wire and this produces a varying magnetic field around the coil.

The varying magnetic field of the coil and the magnetic field of the permanent magnet interact with one another, leading to a varying force being exerted on the cone.

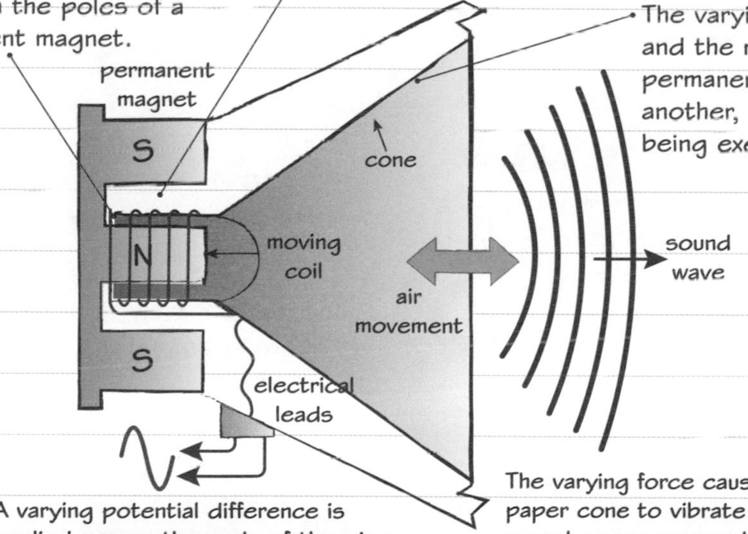

A varying potential difference is applied across the ends of the wire.

The varying force causes the paper cone to vibrate and sound waves are produced.

The **pitch** of the sound from a loudspeaker depends on the **frequency** of vibration of the cone, and the **loudness** of the sound depends on the **amplitude** of its vibration.

Worked example

Explain how a microphone is like a generator and a loudspeaker is like a motor **(4 marks)**

A generator transfers movement, or kinetic energy, into electrical energy when a wire is moved in the region of a magnetic field. This is also what a microphone does.

A motor transfers electrical energy into kinetic energy. The loudspeaker does this by transferring electrical energy into the movement of the cone. This makes use of the magnetic property of electric current, which interacts with the magnetic field of a permanent magnet, causing a force.

Now try this

1 Explain how
(a) a microphone **(2 marks)**
and (b) a loudspeaker **(2 marks)**
operate over the range of audible frequencies.

2 Draw a flow diagram to show the energy stores and transfers for a microphone. **(4 marks)**

Transformers

Transformers can be used to **change the size** of an **alternating voltage** or **current.**

Transformers

A transformer can be used to change the potential difference of an alternating electricity supply.

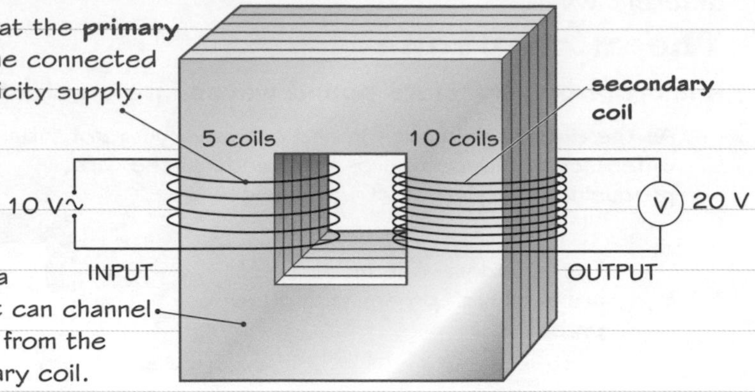
Remember that the **primary coil** is the one connected to the electricity supply.
secondary coil
5 coils
10 coils
10 V~
V
20 V
INPUT
OUTPUT
The core must be made of a magnetic material so that it can channel the magnetic lines of force from the primary coil to the secondary coil.

When an alternating current flows through the primary coil it produces a changing magnetic field in the iron core. This magnetic field induces an alternating potential difference across the secondary coil. The alternating potential difference in the secondary coil leads to an alternating current when the circuit is completed.

Potential difference and turns

The relationship between the potential differences in the primary and secondary coils and the turns ratio is given by the equation:

$$\frac{\text{potential difference across primary coil (V)}}{\text{potential difference across secondary coil (V)}} = \frac{\text{number of turns in primary coil}}{\text{number of turns in secondary coil}}$$

$$\frac{V_p}{V_s} = \frac{N_p}{N_s}$$

because the potential difference across the secondary coil is greater than the potential difference across the primary coil

Worked example

(a) A transformer has 20 turns on its primary coil and 240 turns on its secondary coil. The potential difference across the primary coil is 12 V. Calculate the potential difference across the secondary coil. **(4 marks)**

$$\frac{V_p}{V_s} = \frac{N_p}{N_s}$$

$$V_s = V_p \times \frac{N_s}{N_p}$$

$$= 12\text{V} \times 240 \div 20 = 144\text{V}$$

(b) What type of transformer is this? **(1 mark)**

step-up transformer

(c) Calculate the current in the secondary coil when the current in the primary coil is 18 A. **(4 marks)**

$$V_p \times I_p = V_s \times I_s$$

$$12\text{V} \times 18\text{A} = 144\text{V} \times I_s$$

$$I_s = 12\text{V} \times 18\text{A} \div 144\text{V} = 1.5\text{A}$$

Transformer efficiency

It is assumed that transformers are 100% efficient. This means that the input power of the primary coil and the output power of the secondary coil are equal. They are connected by the equation: $V_p \times I_p = V_s \times I_s$

$$\text{potential difference across primary coil (V)} \times \text{current in primary coil (A)} = \text{potential difference across secondary coil (V)} \times \text{current in secondary coil (A)}$$

Now try this

1 Calculate V_p when $N_p = 40$, $N_s = 800$ and $V_s = 18$ V. **(3 marks)**

2 A transformer with an input potential difference of 30 kV has a power output on the secondary coil of 3 kW. Calculate the current in the primary coil. State the assumption that you make. **(3 marks)**

3 Suggest why transformers will not work with a d.c. input. **(3 marks)**

Transmitting electricity

The **National Grid** is a system of wires that **transmits electricity** from power stations to where the electricity is needed and used.

The National Grid

Electricity is generated and transported to our homes, hospitals and factories by the National Grid. The National Grid is the **wires** and **transformers** that transmit the electricity; it does **not** include the **power stations** and the **consumers**.

To ensure that transmission is efficient, and that very little energy is lost as heat, the voltage and current values need to be chosen carefully at different stages.

Maths skills: Transformer equations

You need to be able to explain the advantages of using high voltages to transmit electrical energy using these four equations:

power = energy ÷ time ($P = E \div t$)

power = current × voltage ($P = I \times V$)

power = current^2 x resistance ($P = I^2 \times R$)

primary voltage × primary current = secondary voltage × secondary current ($V_p \times I_p = V_s \times I_s$)

Transformers and the National Grid

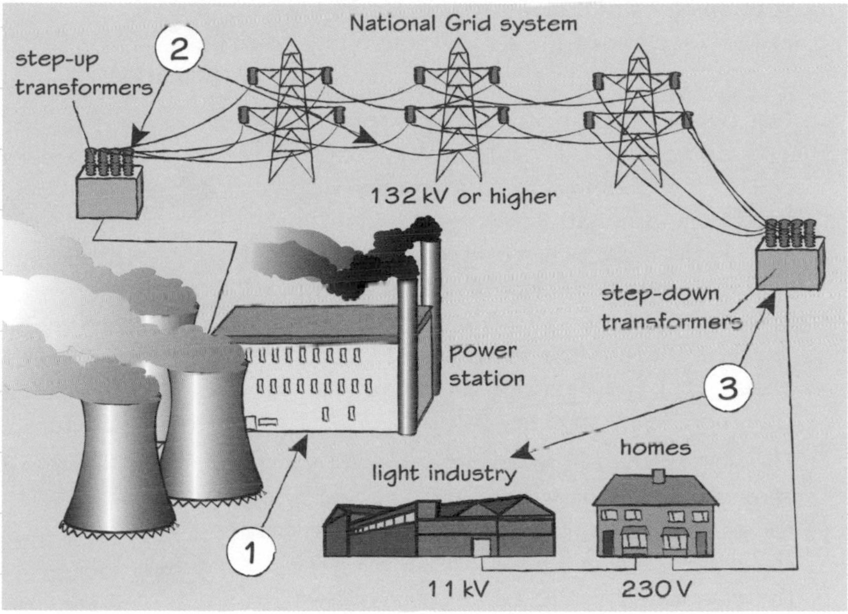

1. **Fossil fuels** or **nuclear fuel** are used to generate electrical energy in the power station.
2. A **step-up transformer** increases the voltage to 132 kV or more. For a fixed amount of electrical energy or power output, increasing the voltage means there will be a decrease in current. Electrical energy is transmitted at a high voltage and low current through the wires to reduce energy losses as heat in the wires. It also means thinner wires can be used, which reduces costs.
3. **Step-down transformers** decrease the voltages from the National Grid for safer use in our homes and industry. Reducing the voltage means there will be an increase in current.

Worked example

Explain why electrical power is transmitted at a high voltage and a low current on the National Grid. **(3 marks)**

Power is transmitted at a high voltage and a low current because the amount of electrical energy wasted as thermal energy is proportional to the square of the current, from the equation $P = I^2R$. So, if we double the size of the current flowing, the power wasted increases by a factor of 2^2 or 4.

Less current means thinner wires. Thinner wires means less wire, so less copper to purchase by the company. Low-resistance wires also reduce power losses.

Now try this

1. Use the Power equations to explain why we do not transmit electrical energy through the National Grid at a high current. **(3 marks)**
2. Describe the best way of transmitting 10^9 W of power over the National Grid. **(2 marks)**

Extended response – Electromagnetic induction

There will be one or more 6 mark questions on your exam paper. For these questions, you will need to think scientifically and structure your answer logically, showing how the points you make are related to each other. You can revise the topics for this question, which is about **electromagnetic induction**, on pages 102–105.

Worked example

Generators are used in the small-scale and large-scale production of electrical energy. The diagram shows an a.c. generator.

Explain how the generator can produce a range of a.c. voltages.

Your answer should refer to the factors that can cause the size or direction of the a.c. output voltage to change.

(6 marks)

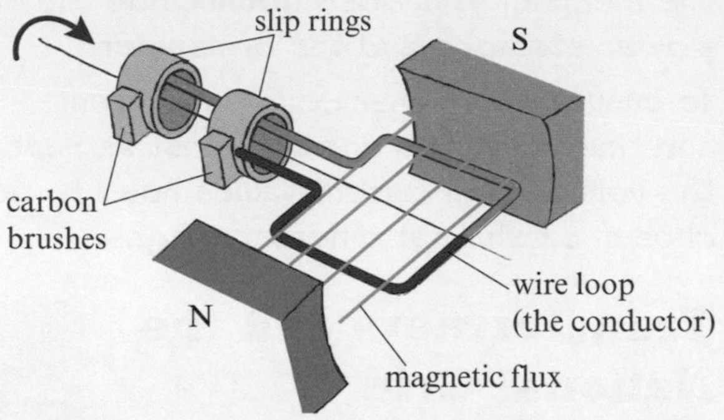

This output is a.c. because the direction of the induced voltage will change every 180° as the coil rotates, leading to an a.c. signal with both positive and negative voltages. When the coil arm is cutting downwards through the lines of magnetic flux, the induced voltage will be in the opposite direction to when the arm is cutting upwards through the lines of flux. The slip rings ensure that the direction of the a.c. output changes direction every half turn.

The size of the output is determined by the strength of the magnetic field, the speed of rotation of the coil and the number of turns on the coil. Increasing any of these will increase the magnitude of the a.c. output voltage.

The direction of the a.c. output is affected by the direction of the magnetic field or flux lines and by the direction of rotation of the coil. Reversing either of these will change the direction of the a.c. output by 180°.

The answer starts by explaining why the output is a.c. in nature, with reference to electromagnetic induction and the structure of the generator.

Command word: Explain

When you are asked to explain something, it is not enough to just state or describe something. Your answer must contain some reasoning or justification of the points you make.

The three main factors that affect the output are stated here. Make sure that you refer to the strength of the magnetic field or magnetic flux and not the size of the magnet. A bigger magnet does not necessarily mean a stronger field.

There are two factors that affect the direction of the a.c. output. Changing either of these factors will reverse the polarity of the output.

It is important to understand the difference between the operation of an electric motor and a generator. Whilst they look very similar, the operation of an electric motor is opposite to that of a generator. The electric motor transfers energy from other stores electrically to kinetic energy, whereas the generator transfers kinetic energy to other stores of energy electrically. Refer to page 101 to compare the two devices.

Now try this

Transformers are used to step up and step down potential difference in the transmission of electricity across the UK. Explain how transformers work and why electricity is transmitted at a large voltage and a small current in the transmission lines of pylons.

Your answer should refer to an equation that relates power loss to electric current. **(6 marks)**

Changes of state

Substances can change from one state of matter to another. The **change of state** depends on whether **energy is gained** or **lost** by the substance.

The three states of matter

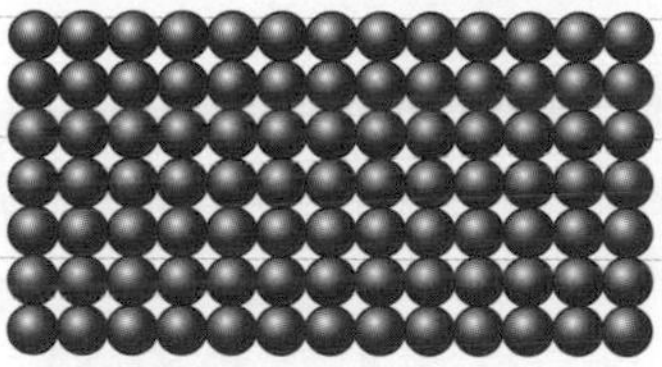

solid

In a solid the particles vibrate but they cannot move freely.

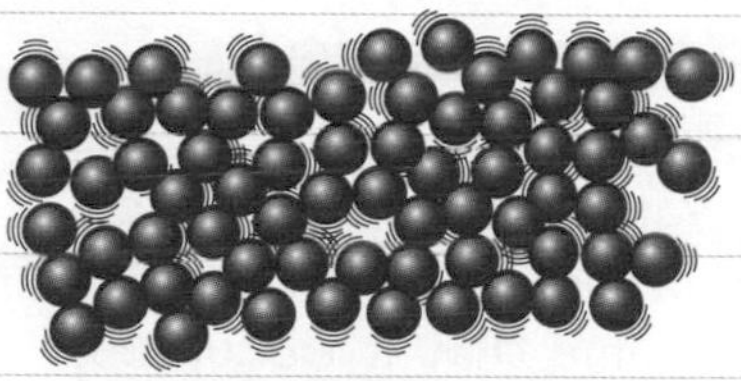

liquid

In a liquid the particles can move past each other and move around randomly.

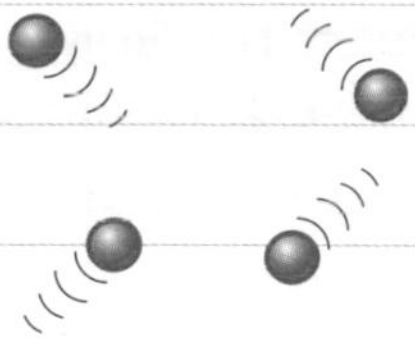

gas

In a gas the particles move around very fast and they move all the time. This is because they have a lot of **kinetic energy**.

Changing state

Changes of state are physical changes rather than chemical changes and they can be reversed.

The mass before the change is equal to the mass after the change. For example, when 500 kg of a solid melts it becomes 500 kg of a liquid.

When a body is heated or cooled, its temperature will change if its state of matter does not change. For example, if 750 ml of a liquid is heated, then the temperature will increase until it reaches its boiling point.

When a material changes state, it does so at a constant temperature. For example, when water boils, 1 kg of water at 100 °C will turn to 1 kg of steam at 100 °C. For more about this, see page 110.

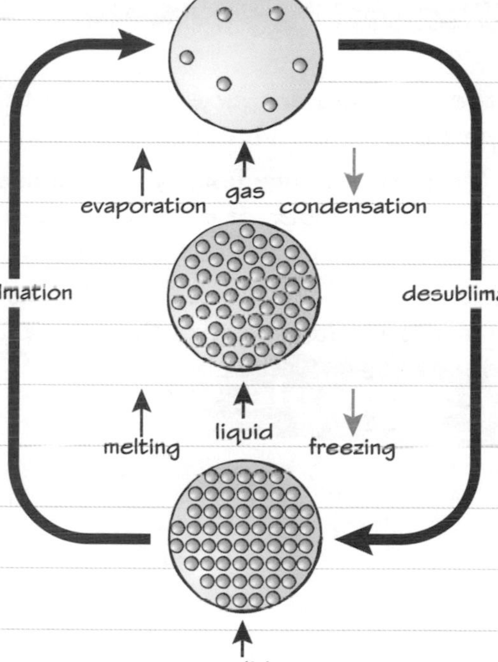

Worked example

Describe what happens when 2.0 kg of steam is allowed to cool against a glass plate, collected and then placed in a freezer. **(4 marks)**

2.0 kg of water will be produced when the steam turns to water at 100 °C. It will then cool as a liquid from 100 °C to room temperature. When placed in the freezer it will cool down from room temperature and turn to ice at 0 °C before continuing to cool to around −18 °C.

Now try this

1 Describe what happens when 1.5 kg of ice is left on a kitchen bench. **(3 marks)**

2 What effect does cooling a gas have on its pressure and volume? **(3 marks)**

3 Describe what happens when 200 kg of solid copper is heated to beyond its boiling point. **(5 marks)**

Density

The **density** of a material is a measure of the **amount of matter** that it contains **per unit volume.**

Density, mass and volume

Changing the amount of material will change both its mass and its volume. If the volume of a block of a particular material is doubled, its mass will also double.

The density of a material does not vary greatly for a given state of matter and relates to how closely packed the atoms or molecules are within the volume that it occupies.

The density of a material in the different states of matter varies greatly because of the arrangement of the particles in the state.

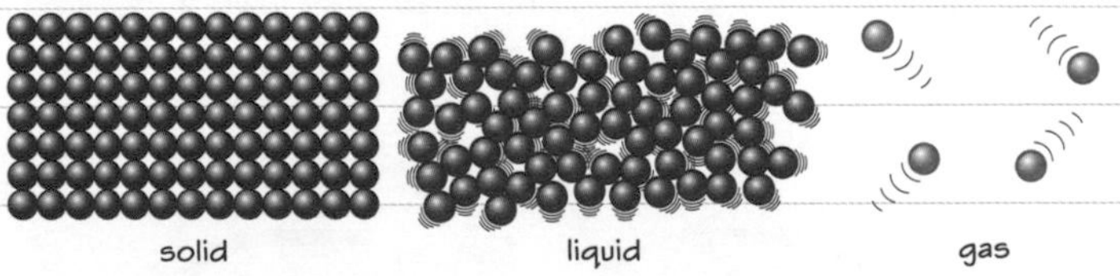

In a solid, the particles are closely packed together. The number of particles in a given volume is high.

In a liquid, the particles are usually less densely packed than in a solid. The number of particles in a given volume is less than for a solid. Solids are high density.

In a gas, the particles are spread out. The number of particles in a given volume is low. Gases are low density.

Calculating density

You can calculate density using the equation:

$$\text{density (kg/m}^3) = \frac{\text{mass (kg)}}{\text{volume (m}^3)}$$

$$\rho = \frac{m}{V}$$

LEARN IT! IT'S NOT ON THE EQUATIONS LIST

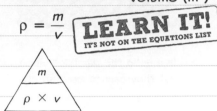

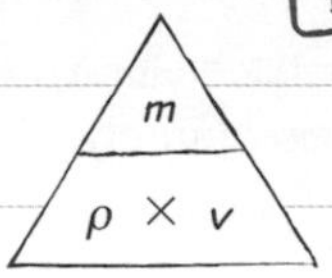

Changing the volume of a substance will change its mass, but its density will be constant.

Other units for density can also be used, such as g/cm³. The density of copper is 8.96 g/cm³. This means that every 1 cm³ of copper metal has a mass of 8.96 g.

Worked example

(a) A body has a mass of 34 000 kg and a volume of 18 m³. Calculate its density. **(4 marks)**

density = mass ÷ volume
= 34 000 kg ÷ 18 m³
= 1900 kg/m³

(b) Calculate the mass of 0.2 m³ of this material. **(4 marks)**

mass = density × volume
= 1900 kg/m³ × 0.2 m³ = 380 kg

Maths skills: Using correct units

When performing calculations involving density, mass and volume check that:

- you are using the correct equation to find density, mass or volume.
- the units are consistent.

When the mass is in kg and the volume is in m³ then the density will be in kg/m³. Sometimes you may be dealing with g and cm³, so check the units.

Worked example

A solid has a density of 3.6 g/cm³ and a mass of 320 g. Calculate its volume. **(4 marks)**

volume = mass ÷ density
= 320 g ÷ 3.6 g/cm³ = 88.9 cm³

Maths skills: Checking units

Since volume = mass ÷ density then the units here will be g ÷ g/cm³. This is the same as g × cm³/g, which means the grams cancel and we are left with cm³ – a unit of volume.

Now try this

1 Calculate the density of a material that has a mass of 862 g and a volume of 765 cm³. **(3 marks)**

2 The density of aluminium is 2700 kg/m³. Calculate the mass of a block of aluminium with a volume of 50 000 cm³. **(4 marks)**

Investigating density

Practical skills You can measure the volume and mass of solids and liquids to determine their densities.

Core practical

Aim

To determine the density of solids and liquids.

Apparatus

measuring cylinder, displacement can, electronic balance, ruler, various solids and liquids

Method 1 (for a solid)

1 Measure the mass of the solid using an electronic balance and record its mass in your results table.

2 Determine the volume of the solid. This can be done by measuring its dimensions and using a formula, or by using a displacement can to see how much liquid it displaces. Record this value in your results table.

Method 2 (for a liquid)

1 Find the mass of the liquid by placing the measuring cylinder on the scales, and then zeroing the scales with no liquid present in the measuring cylinder. Add the liquid to the desired level.

2 Record the mass of the liquid in g from the balance and its volume in cm^3 from the measuring cylinder.

Be careful to read the volume value correctly – at the bottom of the meniscus.

Results

Your results can be recorded in a table.

Material	Mass (g)	Volume (cm^3)	Density (g/cm^3)

3 Find the density of the solid by using the equation density = mass ÷ volume

Conclusion

The density of a solid and a liquid can be determined by finding their respective masses and volumes. Dividing the mass by the volume gives you the density of the solid or liquid.

There is more information about the relationship between mass, volume and density on page 108.

Be careful to use solids and liquids that are safe to use in this investigation. Mercury (a liquid metal) was used in these investigations until it was found to be carcinogenic.

When taking volume readings, make sure that you read the scales with your eye at the same level as the meniscus. Otherwise a parallax error will arise and your values will be incorrect.

Maths skills **Converting between units**

Density can be given in g/cm^3 or kg/m^3.

Since 1 kg = 1000 g and $1\ m^3 = 1\,000\,000\ cm^3$:

- to convert from g/cm^3 to kg/m^3 multiply by 1000
- to convert from kg/m^3 to g/cm^3 divide by 1000.

For example, water has a density of $1\ g/cm^3$ or $1000\ kg/m^3$.

Now try this

1 A measuring cylinder is placed on an electronic balance and its display is set to zero. The measuring cylinder is then filled with a liquid up to a volume of $256\ cm^3$. The mass shown is 454 g. Calculate the density of the liquid. **(3 marks)**

2 Explain how misreading the mass and volume values may occur and how this can lead to a higher than normal density value. **(4 marks)**

Energy and changes of state

You can calculate how much energy you need to **change the temperature** and **state** of a substance.

Specific heat capacity

Specific heat capacity is the **thermal energy** that must be **transferred** to change the **temperature** of **1 kg** of a material **by 1 °C**. Different materials have different specific heat capacities. Water has a value of 4200 J/kg °C, which means that 1 kg of water needs 4200 J of thermal energy to be transferred to raise its temperature by 1 °C.

You can calculate the thermal energy required using the equation: $\Delta Q = m \times c \times \Delta\theta$

change in thermal energy (J) = mass (kg) × specific heat capacity (J/kg °C) × change in temperature (°C)

Specific latent heat

Specific latent heat is the **energy** that must be transferred to change **1 kg** of a material from one **state of matter to another**. There are usually two values for specific latent heat:

1. the **specific latent heat of fusion** when the change of state is between a solid and a liquid – during **melting** or **freezing**
2. the **specific latent heat of vaporisation** when the change of state is between a liquid and a gas – during **boiling** or **condensation**.

You can calculate the thermal energy required using the equation:

thermal energy for a change of state (J) = mass (kg) × specific latent heat (J/kg)

$$Q = m \times L$$

Specific latent heat should **not** be confused with specific heat capacity. Calculations involving **specific latent heat** involve changes of state and **never involve a change in temperature**, since **changes of state always occur** at a **constant temperature**.

Worked example

(a) A mass of 800 g has a specific heat capacity of 900 J/kg °C and is heated from 20 °C to 80 °C. Calculate how much thermal energy was supplied. **(3 marks)**

$\Delta Q = m \times c \times \Delta\theta$

$0.8 \times 900 \times (80 - 20) = 43\,200$ J

(b) Calculate the energy required to convert 40 kg of ice at 0 °C to liquid water at 0 °C. The specific latent heat of fusion of water is 334 000 J/kg. **(3 marks)**

$Q = m \times L$

$Q = 40 \times 334\,000 = 13\,360\,000$ J

Worked example

Ice at –40 °C is heated until it becomes steam at 110 °C. Sketch a graph to show the changes that take place during this process.

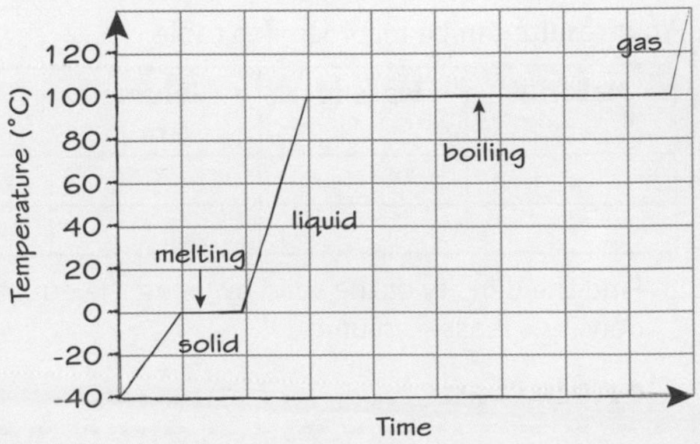

Now try this

1 Calculate the energy required to raise the temperature of 15 kg of water from 18 °C to 74 °C. **(3 marks)**

2 A mass of 1250 g of metal is connected to a heater rated as 2.8 A and 16 V and heated for 12 minutes. The temperature of the metal block increases by 26 °C. Calculate its specific heat capacity. **(4 marks)**

See page 89 for the equation to work out the energy transferred by the heater.

Thermal properties of water

Practical skills You can determine the specific heat capacity of water using an electric heater. You can also observe how water behaves when a sample of ice is melted.

Core practical

Aim

To determine the specific heat capacity of water and to describe the behaviour of ice when melting.

Apparatus

water, ice, thermometers, electric heater, power supply, insulation, beakers, electronic balance

Method

1 Set the apparatus up as shown in the diagram.

2 Measure the mass (m) of the water using an electronic balance.

3 Record the potential difference (V) of the power supply and the current (I) through the heater.

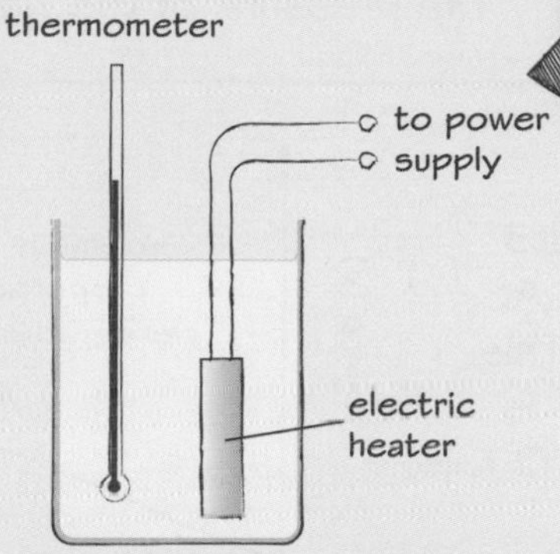

4 Take temperature readings every 30 seconds for a time (t) until the water reaches the desired temperature. Record the initial and final temperatures of the water and find the change in temperature (ΔT).

Results

Plot a graph of temperature against time. Calculate the specific heat capacity. The value c is found by substituting the results into the equation $c = (V \times I \times t) / (m \times \Delta T)$.

Conclusion

The specific heat capacity of water is the energy required to change 1 kg of water by a temperature of 1 °C. Its value is about 4200 J/kg °C. Water melts when it changes from ice to liquid water. There is no change in temperature and 1 kg of ice requires 334 000 J to melt.

There is more information about specific heat capacity on page 110, and more information about the energy transferred by a known current and potential difference on page 89.

Having insulation around the beaker will transfer less energy to the surroundings and give a more accurate value for the specific heat capacity of the water.

When recording the temperature readings, avoid parallax errors by reading the thermometer scale at eye level. Record the temperature values regularly, at equal intervals and ensure that the thermometer is in the middle of the liquid.

A graph of temperature against time for ice melting to water is shown here.

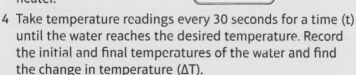

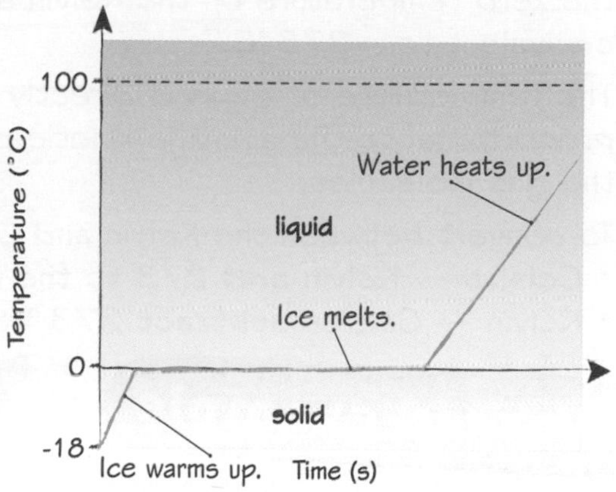

The word 'specific' here in this context is for 1 kg of mass. In other words, the values for specific heat capacity and latent heat are for when 1 kg either changes temperature by 1 °C or when it changes state.

Now try this

1 Explain why insulating the beaker leads to a more accurate value for the specific heat capacity of water. **(3 marks)**

2 Explain why melting occurs without there being any change in temperature. **(3 marks)**

3 Draw the apparatus that would allow you to obtain the temperature–time graph of melting ice. **(3 marks)**

Pressure and temperature

The **pressure** of a **gas** can be explained in terms of the motion of its particles.

The pressure of a gas

Gas pressure depends on the motion of the particles in the gas. Gas particles strike the walls of a container at many different angles. So the pressure of a gas produces a net force at right angles to the wall.

Pressure can be increased by:

- **increasing** the **temperature** of the gas
- **increasing** the **mass** of the gas
- **decreasing** the **volume** of the gas.

Pressure and temperature

The **pressure** of a **fixed mass** of **gas** at a **constant volume** depends on the **temperature** of the gas. When the temperature increases:

1. The gas molecules have a greater average kinetic energy.
2. The gas molecules move faster.
3. There are more collisions between the molecules and the walls of the container each second.
4. More force is exerted on the same area each second.
5. The pressure of the gas increases.

Absolute zero

As a gas is cooled the average speed of its particles falls and its volume gets smaller. At −273 °C the gas volume shrinks to zero. This temperature is called **absolute zero**. This is the zero temperature on the Kelvin scale. 0 K is equivalent to −273 °C.

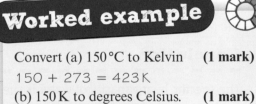

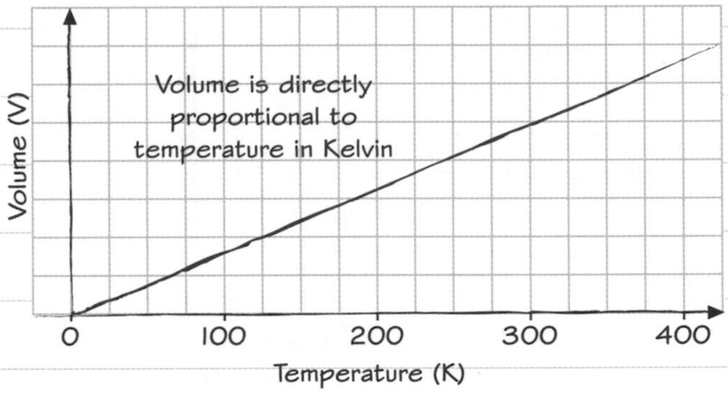

The temperature of a gas is directly proportional to the average kinetic energy of the gas molecules.

To convert between the Kelvin and Celsius scales use the appropriate formula:

- Celsius → Kelvin **add** 273 to the Celsius temperature
- Kelvin → Celsius **subtract** 273 from the Kelvin temperature

Worked example

Explain gas pressure using kinetic theory. **(3 marks)**

The particles (atoms or molecules) in a gas are continuously moving in a random way and colliding with the container walls. The force from these collisions produces pressure on the walls. On average the number and force of collisions is the same in all directions, so pressure is the same on all the walls of the container.

Worked example

Convert (a) 150 °C to Kelvin **(1 mark)**

150 + 273 = 423 K

(b) 150 K to degrees Celsius. **(1 mark)**

150 − 273 = −123 °C

Now try this

1 (a) Convert 20 °C to a temperature on the Kelvin scale. **(1 mark)**
(b) Convert 300 K to its equivalent temperature on the Celsius scale. **(1 mark)**

2 The temperature of a fixed amount of gas is increased from −23 °C to 227 °C. Explain what happens to the average kinetic energy of the particles in this gas. **(3 marks)**

3 (a) State what happens to the average speed of gas particles as a gas is cooled. **(1 mark)**
(b) Explain the effect this has on the pressure of the gas in a rigid sealed container. **(2 marks)**

Volume and pressure

The **change in the volume** of a gas is **inversely proportional** to the change in pressure when the mass and temperature are constant.

Changing the volume of a gas at constant temperature

Pressure and **volume** are **inversely proportional** to one another.

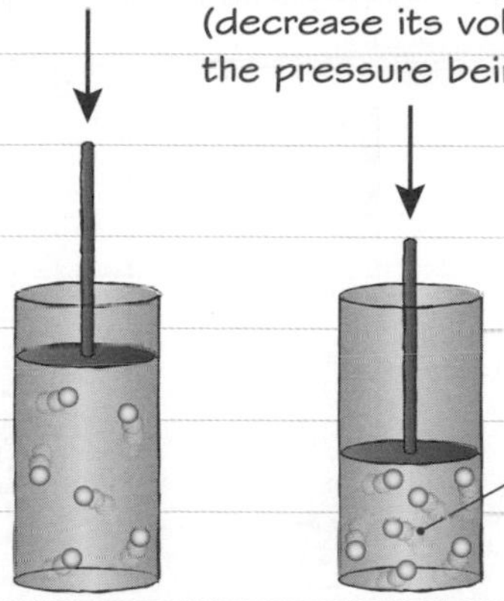
You can make a gas contract (decrease its volume) by increasing the pressure being exerted.
The product of pressure and volume, P × V, will always have the same value.
Decreasing the volume of a gas at a constant temperature means that the moving gas particles will collide more frequently with the walls of the container they are in. The pressure of the gas will increase.
10 m³ × 100 kPa = 1 000 000 Pa
5 m³ × 200 kPa = 1 000 000 Pa
As the pressure increases then the volume of the gas must decrease if the temperature of the gas is kept constant.

Gas equation

You can use this equation to calculate the pressure or volume of gases for a fixed mass of gas at a constant temperature:

$P_1 \times V_1 = P_2 \times V_2$

The SI units for *P* are pascals (Pa) and for *V* are m^3.

1 kPa = 1000 Pa

The units for pressure and volume do not have to be in SI units as long as the same units are used throughout.

See page 119 for more on pressure.

Worked example

At atmospheric pressure, 100 kPa, the volume of the column of trapped gas in the apparatus shown is 28 cm³. The pump is used to increase the pressure on the trapped gas to 250 kPa. Calculate the new volume of the trapped air. **(4 marks)**

Rearrange the equation given on the left to give:

$$V_2 = \frac{V_1 \times P_1}{P_2}$$

where V_1 = 28 cm³, P_1 = 100 kPa and P_2 = 250 kPa

$$V_2 = \frac{28\,\text{cm}^3 \times 100\,\text{kPa}}{250\,\text{kPa}} = 11.2\,\text{cm}^3$$

Doing work on a gas

When a gas is compressed quickly, its volume decreases quickly and its temperature can rise. This is because work is being done in compressing the gas. This leads to an increase in the temperature of the gas due to the work causing an increase in the average kinetic energy of the particles in the gas. One such example is the warming up of the air inside a bicycle pump when it is used to inflate a tyre.

Now try this

See page 118 for more on the pressure in a fluid.

1 If P_1 = 200 kPa and V_1 = 28 cm³, calculate the new pressure P_2 when the volume is increased to 44.8 cm³. **(3 marks)**

2 Explain how it is possible for a gas to be
(a) compressed at a constant temperature
(b) compressed so that its temperature increases. **(4 marks)**

3 The pressure in a liquid increases with depth. A bubble of carbon dioxide forms in a glass of fizzy drink; explain what you would expect to observe, stating any assumptions you make. **(4 marks)**

Extended response – Density

There will be one or more 6 mark questions on your exam paper. For these questions, you will need to think scientifically and structure your answer logically, showing how the points you make are related to each other. You can revise the topics for this question, which is about the **particle model**, on pages 107–113.

Worked example

A team of investigators has been asked to determine whether rings being sold in a jeweller's shop are made from pure gold or whether they are fake.
Explain how the investigators could determine whether the rings were made from pure gold, which has a density of 19.3 g/cm^3.
Your answer should refer to a suitable experimental method that will allow you to determine the density of the material and any equations you would need use. **(6 marks)**

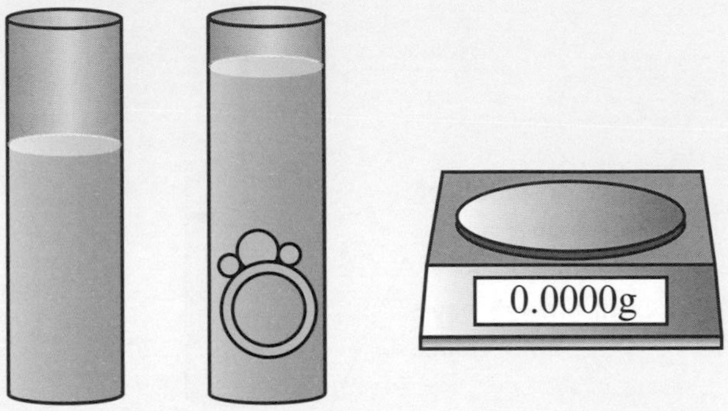

Density = mass ÷ volume; it is the mass per unit volume of a material. It is fixed for a substance in a given phase of matter.

The answer starts with a brief explanation of what density is, as well as stating the equation.

To find the mass of a ring, we would put it on the electronic balance and get its mass in grams. To find the volume of the ring, we would lower it into a measuring cylinder of water and find the difference in the volume readings on the measuring cylinder scale. We do this by subtracting the volume reading before the ring has been submerged in the water from the volume reading after the ring has been submerged. This difference in volume is equal to the volume of the ring in cm^3. Dividing the value for the mass of the ring in g by the volume of the ring in cm^3 will get the value for the density of the ring in g/cm^3.

The method addresses how the mass and volume can be calculated. You could also refer to the use of a displacement can to determine the ring's volume.

If the value obtained for the density is very close to 19.3 g/cm^3 then the ring *might be* made from gold. If the density is very different from the value of 19.3 g/cm^3 then it is not pure gold.

The final part of the answer explains how the density value obtained tells you whether the material is gold or not – notice how the answer states that the ring *might be* gold, since you cannot say for definite as other materials may have this density.

When you structure your answer, make sure that you address the question in an order that allows you to deal with each of the points being asked for. Do not just list equations and facts. Communicate to the examiner that you understand how to find the density of the rings and how it will allow you determine if the rings are real or fake based on the known value for the density of gold.

Now try this

A gas contains many particles moving with random motion in a container of a fixed volume.
Explain how the pressure of a gas can be increased by changing its volume or temperature.
Your answer should refer to energy, forces and the motion of the particles in the gas. **(6 marks)**

Elastic and inelastic distortion

The distortion of a material may be described as being **elastic** or **inelastic** and **requires more than one force**.

Bending, stretching and compressing

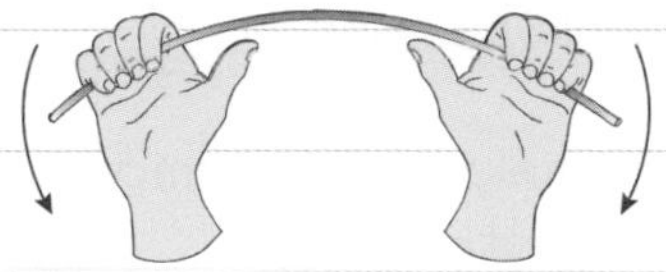

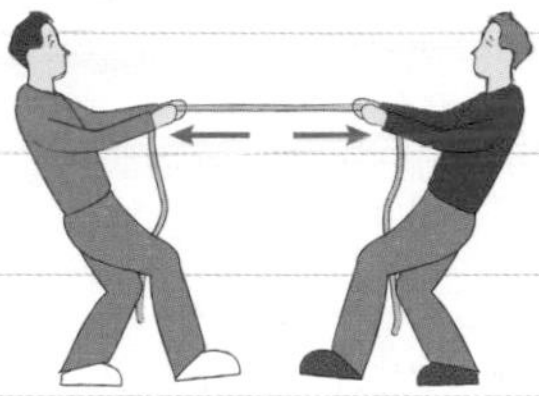

Bending requires two forces, one acting clockwise and one acting anticlockwise.

Stretching requires two forces of tension, acting away from each other.

Compression involves two equal forces acting towards one another.

Elastic distortion

Elastic distortion means that a material will return to its original shape when the deforming force is removed.

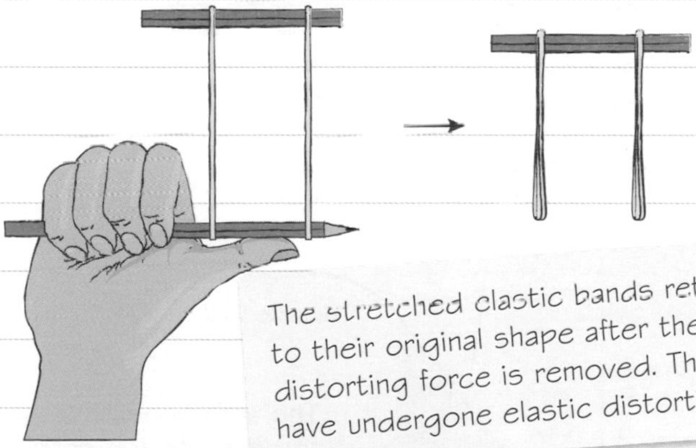

The stretched elastic bands return to their original shape after the distorting force is removed. They have undergone elastic distortion.

Inelastic distortion

Inelastic distortion means that a material will not return to its original shape when the deforming force is removed.

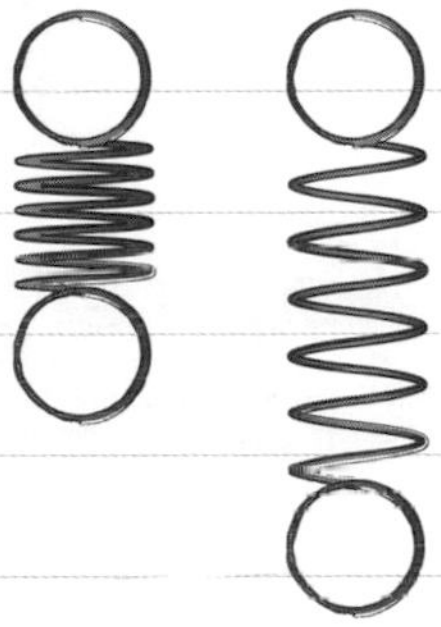

This spring has not returned to its original shape after being distorted, so it has undergone inelastic distortion.

Worked example

Describe what this graph shows for a spring that has been stretched. **(3 marks)**

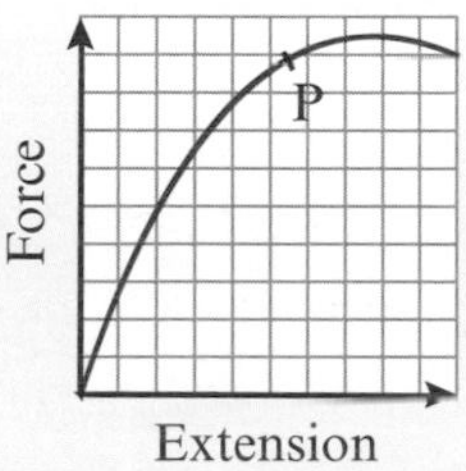

Initially, the relationship between force and extension is linear, so the behaviour of the spring is elastic for smaller forces.

Point P is the elastic limit. For forces applied after this, the relationship is no longer linear and the spring will not return to its original shape – it will exhibit inelastic distortion.

Now try this

1 Explain the types of forces that cause
 (a) stretching **(2 marks)**
 (b) compression. **(2 marks)**

2 Give examples of materials that exhibit
 (a) elastic distortion **(2 marks)**
 (b) inelastic distortion. **(2 marks)**

3 Explain why materials that are distorted often do not return to their original shape. **(3 marks)**

Springs

Forces can lead to the **distortion** of **elastic** objects, resulting in **energy** being stored.

Elastic distortion

A force applied to a spring can make it undergo linear elastic distortion. The equation that describes this is:

force exerted on a spring (N) = spring constant (N/m) × extension (m)

$F = k \times x$ **LEARN IT!** IT'S NOT ON THE EQUATIONS LIST

You can calculate the work done in stretching a spring using the equation:

energy transferred in stretching (J) = 0.5 × spring constant (N/m) × extension^2 (m^2)

$E = \frac{1}{2} \times k \times x^2$

Force and linear extension

For elastic distortion, **extension** is **directly proportional** to the **force** exerted, **up to the elastic limit.**

The extension is the amount the spring stretches.

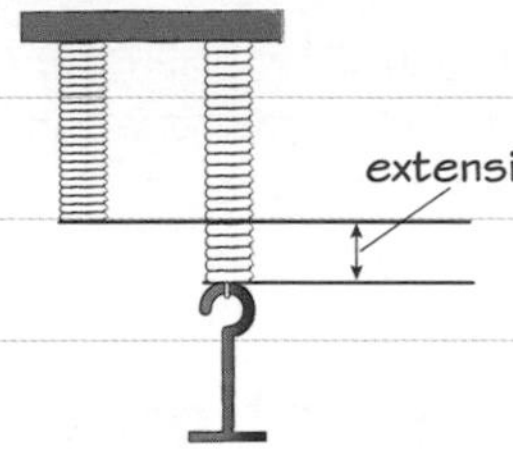

extension, x

Worked example

A spring is distorted elastically. It increases its length by 25 cm when a total weight of 12 N is added.

Calculate

(a) the spring constant of the spring **(3 marks)**

$k = F \div x = 12\,\text{N}/0.25\,\text{m} = 48\,\text{N/m}$

(b) the total energy stored in the spring. **(4 marks)**

energy stored = work done

$E = \frac{1}{2}k \times x^2$

$= \frac{1}{2} \times 48\,\text{N/m} \times (0.25\,\text{m})^2$

$= 1.5\,\text{J}$

When calculating the spring constant, work done or energy stored, ensure that force is in N and extension is in m.

Always use the extension. You may need to calculate this:

extension = total stretched length − original length

Force–extension graph

The gradient tells you the value of the spring constant, k. This is only true when the spring is showing elastic behaviour, i.e. when the line is straight.

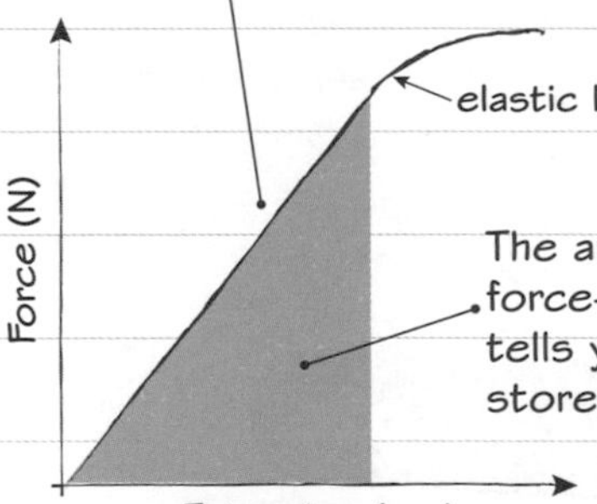

elastic limit
The area beneath a force-extension graph tells you the energy stored in the spring.
Force (N)
Extension (mm)

The area under the graph is the area of a triangle $= \frac{1}{2} \times F \times x$

The force is given by the equation $F = k \times x$

Substituting this into the equation gives

$E = \frac{1}{2} \times k \times x \times x = \frac{1}{2}k \times x^2$

Now try this

1 (a) Calculate the force that is required to extend a spring of spring constant 30 N/m by 20 cm. **(3 marks)**

(b) Calculate how much work is done on the spring under these conditions. **(3 marks)**

2 Explain how the stiffness of a spring is related to its spring constant. **(3 marks)**

3 Describe what happens to a spring for it to start displaying inelastic distortion. **(3 marks)**

Forces and springs

Practical skills It is possible to determine the extension of a spring, and the work done, by applying different forces to the spring and then plotting a graph.

Core practical

Aim

To determine the extension and work done when applying forces to a spring.

Apparatus

spring, ruler, weights, retort stand, boss and clamp

Method

Arrange the apparatus as shown, with the ruler placed vertically alongside the spring and parallel to it. Add masses or weights and collect enough values to plot a graph of force against extension.

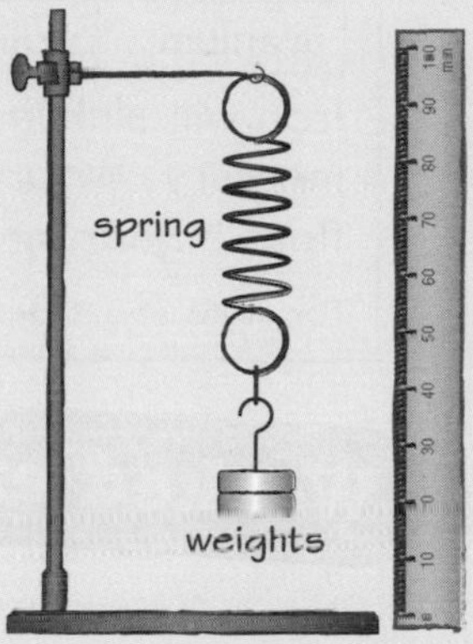

Wear eye protection when taking readings. A stretched spring stores a lot of elastic energy and this could damage your eyes if released.

When taking readings of the extension, read the values on the ruler at eye level to avoid parallax errors.

Results

Taking readings for force and extension allows a force–extension graph to be plotted:

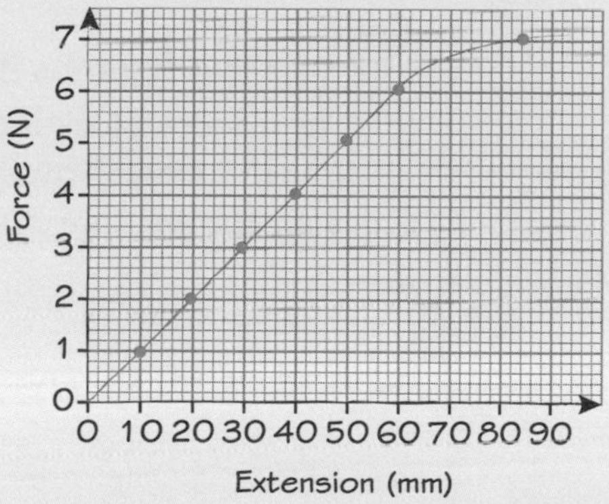

Conclusion

The gradient of the linear part of a force–extension graph gives you the value for the spring constant of the spring when it behaves elastically. The area beneath the graph equals the work done or the energy stored in the spring as elastic potential energy.

Weight and mass

Weight is a force, not a mass. A mass of 100 g is equivalent to a weight or force of 1 N.

Extension is equal to

new length (m) − original length (m).

Work done is equal to the energy stored in the spring.

The gradient of a force–extension graph gives you the spring constant.

The area beneath the line is the work done, or the energy stored as elastic potential energy, by the spring.

Maths skills You will need to convert extension values from mm to m before you can calculate an energy value, and state the result in joules, J. Forces also need to be stated in N, so any mass values recorded need to be converted from g to N.

Now try this

1 Calculate the energy stored by a spring of spring constant 0.8 N/m when extended by 48 cm. **(3 marks)**

2 Explain whether the spring constant and energy stored can be calculated for springs that are being compressed. **(3 marks)**

Upthrust and pressure

The **upthrust**, **weight** and **density** of a fluid will determine whether an object will **sink** or **float** in it.

Upthrust

A body floating in a fluid is subject to an upwards force called the upthrust. This force is equal to the weight of the fluid that is displaced.

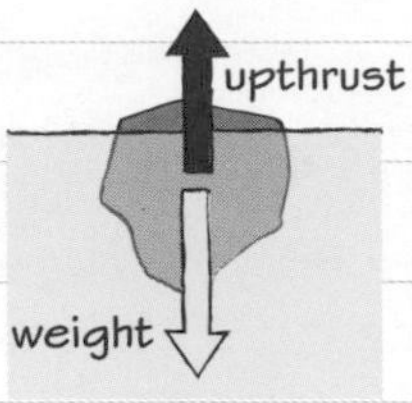

This is why a small weight will float higher up in water than a large weight.

A **fluid** is a liquid or a gas. Water and air are both examples of fluids.

Pressure

Pressure is the force acting per unit area, measured at right angles to the area. Pressure is measured in pascals (Pa): 1 Pa = 1 N/m^2.

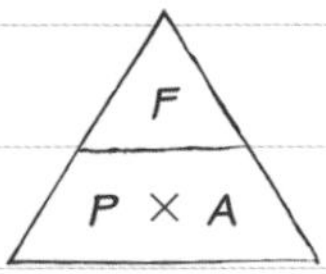

You can use this equation to calculate pressure:

pressure (Pa) = force normal to surface (N) ÷ area of surface (m^2)

$$P = \frac{F}{A}$$

LEARN IT! IT'S NOT ON THE EQUATIONS LIST

A high force over a small area can produce a high pressure. A low force over a large area can produce a low pressure.

Floating and density

A body will float in a fluid if its density is less than that of the fluid. A body will sink in a fluid if its density is greater than that of the liquid.

Object	State	Density
water	liquid	1000 kg/m^3
mercury	liquid	13 600 kg/m^3
iron	solid	7800 kg/m^3
wood	solid	740 kg/m^3
uranium	solid	18 700 kg/m^3

Iron will sink in water but will float in mercury. Wood will float in water and will float higher up in mercury.

For more about density, see page 108.

Worked example

(a) A force of 100 N acts normal to a block of wood of area 0.1 m^2 in contact with the floor. Calculate the pressure exerted. **(3 marks)**

$P = F \div A$
$= 100\,\text{N} \div 0.1\,\text{m}^2 = 1000\,\text{Pa}$

(b) The wooden block is turned on its end. The area in contact with the floor is now 0.01 m^2.
Calculate the pressure exerted now. **(3 marks)**

$P = F \div A$
$= 100\,\text{N} \div 0.01\,\text{m}^2 = 10\,000\,\text{Pa}$

(c) Explain why the pressure has increased. **(3 marks)**

Same force, smaller area, so pressure is greater.

Worked example

Explain why the ship floats higher up in salty water than in fresh water. **(4 marks)**

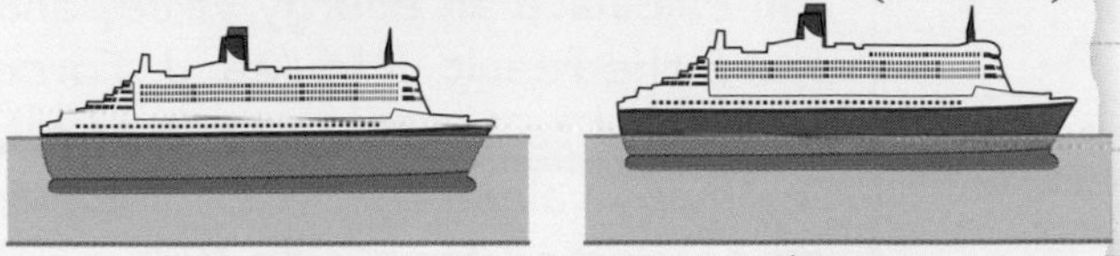

fresh water | salty water

A body will float when the upthrust is equal to its weight. The salty water has a greater density than the fresh water. This means that a lower volume of salty water needs to be displaced to have the same weight as the fresh water that is displaced. Since the area of the ship's bottom is constant, the depth will be less.

Now try this

Refer to density and upthrust in your answer.

1 Calculate the area over which a force of 600 N exerts a pressure of 20 kPa. **(4 marks)**

2 The diagrams show three objects in water. Explain what you can determine about their densities and the mass of water that they displace in each case. **(6 marks)**

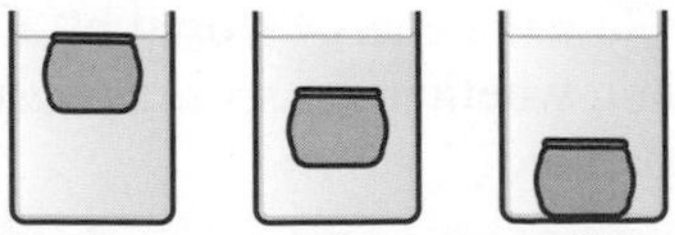

A B C

Pressure and fluids

Fluids (liquids and gases) exert **pressure**, which depends on the **depth** and **density** of the fluid.

Pressure, depth and density

The pressure exerted in a liquid will increase with depth. This is because the further down you go in the liquid, the greater the weight of liquid there will be above.

The pressure in a fluid causes a force that is normal to any surface. This means that the force is at right angles or perpendicular to the surface on which it acts.

You can calculate the pressure in a liquid using the equation:

pressure due to a column of liquid (Pa) = height of column (m) × density of liquid (kg/m³) × gravitational field strength (N/kg)

$$P = h \times \rho \times g$$

The total pressure acting in a fluid depends on the pressure from the fluid and due to the atmospheric pressure.

Atmospheric pressure

The atmosphere can be thought of as a very high column of air above our bodies. This means that it will exert a pressure due to its weight, which acts normal to an area, A. Atmospheric pressure is about 100 000 Pa.

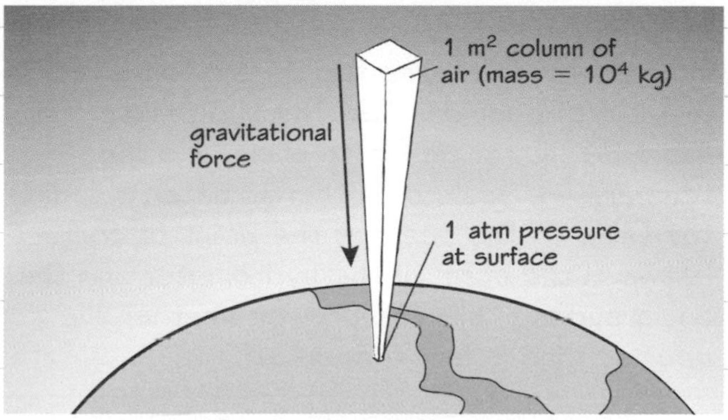

The further up you go in the Earth's atmosphere, the less air there will be above you in the column of air – this means there is less weight of air above you, and hence less pressure, the higher up you go.

Pressure and density

The pressure exerted at a given depth in a liquid depends on the density of the liquid. A greater density means more weight will be acting downwards from above that depth.

The pressure exerted by the water is greater than the pressure exerted by the cooking oil from the same height, h, because the water has a higher density.

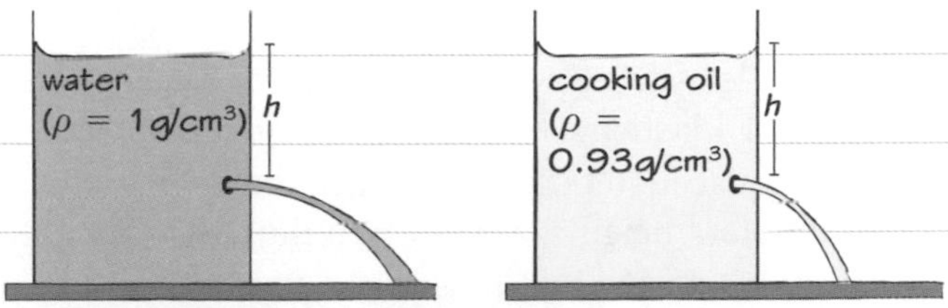

Worked example

(a) Water has a density of 1000 kg/m³. Assume g = 10 N/kg on Earth. Calculate the pressure exerted in water at a depth of 18 m due to the water. **(3 marks)**

$P = h \times \rho \times g$
= 18 m × 1000 kg/m³ × 10 N/kg
= 180 000 Pa

(b) Calculate the total pressure acting due to the atmosphere and the fluid at a depth of 18 m. **(3 marks)**

total pressure = pressure from fluid + atmospheric pressure
= 180 000 Pa + 100 000 Pa
= 280 000 Pa

(c) Work out the total pressure at a depth of 18 m in mercury (density = 13 600 kg/m³). **(4 marks)**

total pressure = pressure from mercury + atmospheric pressure
= 18 m × 13 600 kg/m³ × 10 N/kg + 100 000 Pa
= 2 448 000 Pa + 100 000 Pa
= 2 548 000 Pa

Now try this

You will also need to use the formula for volume.

1 State how pressure changes with
(a) depth **(1 mark)**
(b) density. **(1 mark)**

2 Calculate the pressure acting at a depth of 760 mm in mercury. Ignore atmospheric pressure. **(3 marks)**

3 Show how the equations $P = F \div A$, $\rho = m \div V$ and $W = m \times g$ lead to $P = h \times \rho \times g$. **(4 marks)**

Extended response – Forces and matter

There will be one or more 6 mark questions on your exam paper. For these questions, you will need to think scientifically and structure your answer logically, showing how the points you make are related to each other. You can revise the topics for this question, which is about **forces and matter**, on pages 115–119.

Worked example

A hydrometer is a piece of equipment that is used to determine the density of a liquid.

A simple hydrometer can be made by placing a small mass of Plasticine on one end of a drinking straw.

When placed in different liquids, the hydrometer will float at different heights, as shown in the diagram.

Explain why the hydrometer floats at different heights in the three liquids.

Your answer should refer to the terms density, upthrust and weight. **(6 marks)**

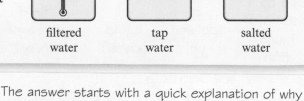

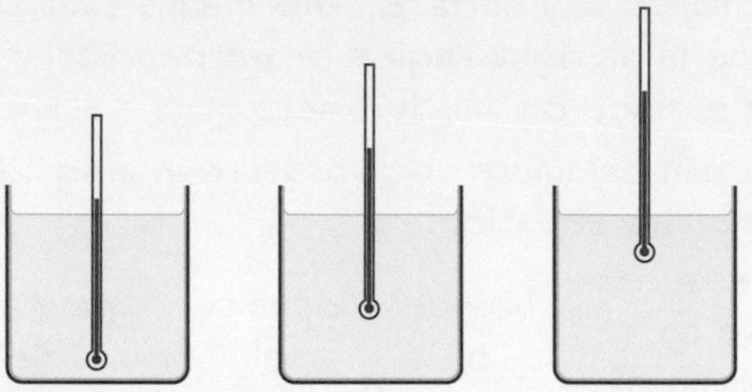

When a body is lowered into a liquid it will float or sink. The body will sink if its density is greater than that of the liquid: the body will sink if it displaces a weight of water that is less than its own weight.

When a body floats in a liquid, it displaces a weight of water that is equal to its own weight. This weight of water is what provides the upthrust to keep the body floating. If the density of the liquid that the body floats in is increased, then less volume of the liquid has to be displaced to provide the same upthrust. Since the base area of the hydrometer is constant, the only variable that needs to be considered will be the height at which the body floats. A body will float higher up (less of it submerged) in a liquid of higher density than in a liquid of lower density.

This means that the salted water is the most dense of the liquids since the hydrometer is floating higher up in that than in the other liquids. By the same argument, the filtered water has the lowest density of the three liquids as the hydrometer floats lowest down in this liquid.

The answer starts with a quick explanation of why objects sink or float in a liquid.

Weight, upthrust and density are introduced, as asked for in the question. The relevance of these three terms is then related to the context and an explanation of the height at which the hydrometers float is then provided.

Hydrometers

Hydrometers are used in the brewing industry to determine the density or 'specific gravity' of beer or wine. The hydrometer has a density scale on its stem, like the temperature scale on a thermometer. Reading the value on the scale at the point of contact between the stem of the hydrometer and the liquid surface tells the brewer whether the beer or wine is fully fermented.

Having explained why density is an important factor, and how it affects the height at which the hydrometer floats in the liquid, the order of densities of the liquids is then explained.

Now try this

Describe the difference between elastic and inelastic distortion and explain how the spring constant of a spring and the elastic potential energy it stores can be determined when it is behaving elastically.

Your answer should refer to any graph, or equations, that may be useful. **(6 marks)**

Answers

Extended response questions

Answers to 6-mark questions are indicated with a star (*).

In your exam, your answers to 6-mark questions will be marked on how well you present and organise your response, not just on the scientific content. Your responses should contain most or all of the points given in the answers below, but you should also make sure that you show how the points link to each other, and structure your response in a clear and logical way.

1. Key concepts

1 (a) 3 **(1)** (b) 4 **(1)** (c) 5 **(1)**

2 (a) 12 × 60 × 60 **(1)** = 43 200 s **(1)**
(b) = 17.8 m/s **(1)**

3 58.3 MW = 58 300 000 W **(1)** = 5.83 × 10^7 W **(1)**

4 186 000 miles/s = 186 000 × 1609 **(1)** = 299 274 000 m/s **(1)** = 2.993 × 10^8 m/s **(1)**

2. Scalars and vectors

1 (a) mass **(1)** (b) electric field **(1)**

2 (a) 4 m/s **(1)** (b) –4 m/s **(1)**

3 The satellite travels at a constant speed as it covers an equal distance per second **(1)**, but its direction is constantly changing as it moves in a circle **(1)** so its velocity must be changing as velocity is a vector quantity with both size and direction **(1)**.

3. Speed, distance and time

1 speed = 120 m ÷ 8 s **(1)**, speed = 15 **(1)** m/s **(1)**

2 gradient of zero at origin **(1)**, increasing gradient of graph shown **(1)**, distance and time shown as correct *y*- and *x*-axes, respectively **(1)**

3 112.7 km = 112 700 m **(1)**; 1 hour = 60 × 60 = 3600 s **(1)**; 112 700 m ÷ 3600 s = 31.3 m/s **(1)**

4. Equations of motion

1 a = (8 m/s – 2 m/s) ÷ 5 s **(1)** = 6 m/s ÷ 5 s = 1.2 **(1)** m/s² **(1)**

2 $v^2 - 0^2$ = 2 × 1.6 m/s² × 1800 m **(1)** = 5760 m²/s; $v = \sqrt{5760}$ = 75.9 **(1)** m/s **(1)**

3 $v^2 - u^2 = 2ax$, (18 m/s)² – (5 m/s)² = 2 × 1.2 m/s² × x **(1)**; $x = (18^2 - 5^2) \div 2.4$ **(1)** = 124.6 **(1)** m **(1)**

5. Velocity/time graphs

1 (a) Gradient of a velocity–time graph gives acceleration. **(1)**
(b) Area beneath a velocity–time graph gives distance travelled **(1)**.

2 (a) correct axes shown: velocity on *y*-axis and time on *x*-axis **(1)**, horizontal line at 8 m/s shown for 12 s **(1)**, positive linear gradient of 1.5 shown for the next 6 seconds **(1)**, final velocity of 17 m/s shown after 18 s **(1)**
(b) distance travelled = area under graph **(1)**, distance travelled in first 8 s = 8 m/s × 12 s = 96 m **(1)**, distance travelled in second part of journey = 8 m/s × 6 s + $\frac{1}{2}$ × 9 m/s × 6 s = 48 m + 27 m = 75 m **(1)**, total distance = 96 m + 75 m = 171 m **(1)**

6. Determining speed

1 time taken **(1)** for the object to travel over a known distance **(1)**

2 Calculate the speed of the ball through the first light gate using speed = distance ÷ time **(1)**. Calculate the speed of the ball through the second light gate **(1)**. Work out the change in speed and the time difference for the two gates **(1)**. Calculate the acceleration using the equation acceleration = change in speed ÷ change in time **(1)**.

3 Answer should include reference to: measurement of a height, *h*, through which a ball will fall vertically **(1)**; the time taken for the ball to fall through the height, *h* **(1)**; the use of the equation $g = 2h/t^2$ **(1)** to find *g*; repeat values being taken **(1)** and an average value being calculated and compared to *g*; the falling object must accelerate and not reach terminal velocity during the fall **(1)**.

7. Newton's first law

1 balanced or equal and opposite **(1)** OR no forces acting **(1)**

2 The forces acting on the moving body are in different/opposite directions **(1)** of different sizes **(1)**, or a diagram drawn to show this. These forces could be antiparallel or even at an angle, such as at 90° to the initial movement. A change in direction will occur if the forces oppose the original motion or act in a way to change the direction.

3 diagrams showing: (a) two forces of 100 N (1) acting in directly opposite directions (1); (b) two forces of 100 N (1) acting in directly opposite directions (1); (c) two forces of 100 N (1) both acting to the left (1)

8. Newton's second law

1 F = 1.2 kg × 8 m/s² **(1)** = 9.6 **(1)** N **(1)**

2 $m = F \div a$ = 18.8 N ÷ 0.8 m/s² **(1)** = 23.5 **(1)** kg **(1)**

3 80 g = 0.08 kg **(1)**, 0.6 kN = 600 N **(1)**, $a = F \div m$ = 600 N ÷ 0.08 kg **(1)** = 7500 **(1)** m/s² **(1)**

9. Weight and mass

1 $g = W \div m$ **(1)** = 54 N ÷ 18 kg **(1)** = 3 N/kg **(1)**

2 Mass is 5 kg **(1)** as mass is not affected by gravitational field strength; weight = 5 kg × 25 N/kg **(1)** = 125 N **(1)**

10. Core practical – Force and acceleration

1 More accurate/repeat readings **(1)** can be taken with little or no human error **(1)** and calculations can be performed quickly by the data logger to provide speeds and accelerations, etc. **(1)**.

2 more accurate mass **(1)**, more accurate slope/gradient **(1)**, reduce friction, more readings for mass **(1)**, repeat readings to determine the precision **(1)**

3 constant mass throughout / other variables controlled **(1)**, change slope by 10° at a time **(1)**, same starting point along slope **(1)**, record enough values for change in velocity and change in times between two light gates **(1)**, repeat readings to determine the precision **(1)**

11. Circular motion

1 (a) arrow pointing straight up labelled thrust from engine **(1)**, arrow pointing to centre of circle labelled centripetal force **(1)**

(b) It changes direction from straight up to towards the right **(1)** but the magnitude stays the same **(1)**.

12. Momentum and force

1 momentum = mass × velocity **(1)** = 25 kg × 6 m/s **(1)** = 150 kg m/s **(1)**

2 F = 1580 kg (16 m/s – 7 m/s) ÷ 0.8 s **(1)** = 17 775 **(1)** N **(1)**

3 Use safety features such as seat belts and crumple zones **(1)** to increase the time taken for the car to decelerate/change in momentum **(1)** so the force needed decreases **(1)**.

13. Newton's third law

1 For two objects involved in a collision, the forces involved are equal in magnitude but opposite in direction **(1)**, and as the time of impact is the same for both bodies, the impulses and the momentum must be equal and opposite **(1)**.

2 momentum before = 12 kg × 8 m/s = 96 kg m/s **(1)**, momentum afterwards = 96 kg m/s **(1)**, velocity afterwards = 96 kg m/s ÷ 20 kg **(1)** = 4.8 m/s **(1)**

14. Human reaction time

1 (a) Human reaction time is the time between a stimulus occurring and a response **(1)**. (b) any three of: tiredness **(1)**, alcohol **(1)**, drugs **(1)**, distractions **(1)**, age **(1)**.

2 Reaction time is proportional to the square root of the distance an object falls **(1)**. The square root of 4 is 2, so four times the distance means twice the reaction time **(1)**.

15. Stopping distance

1 Thinking distance increases by a factor of four **(1)**, braking distance increases by a factor of 16 **(1)**.

2 $F \times d = \frac{1}{2} \times m \times v^2$, d = 1250 kg × (12 m/s)^2 ÷ (2 × 1800 N) **(1)** = 50 **(1)** m **(1)**

16. Extended response – Motion and forces

*Answer could include the following points:

- reaction times and thinking distance. As reaction time increases so does thinking distance. Reaction time is the time between seeing the hazard and applying a force to the brakes.
- speed of car – affecting thinking and braking distance. More speed means a greater distance is covered when thinking and braking compared with a lower speed.
- alcohol, drugs or tiredness affecting reaction times and thinking distance – both will increase if drivers are under the influence of these since they will be less alert.
- condition of tyres – bald tyres means greater braking distance as there will be less grip.
- condition of road (icy or wet) resulting in less grip.
- mass of car/number of passengers increases braking distance at a given speed due to greater inertial mass/constant force needed over a longer time to stop.

17. Energy stores and transfers

1 Energy store diagram should include the chemical energy, kinetic energy and gravitational potential energy stores and the thermal energy store shown as the wasted form. At each part of the journey of the car uphill, the total of these four stores should be constant. The diagram should show

chemical store (decreases) → (mechanical) (energy transfer) → kinetic store (increases)

+ gravitational potential store (increases) + thermal store (increases)

(4)

2 Diagram should show:

chemical store → kinetic store + gravitational potential store → gravitational store + thermal store

The initial energy store is chemical energy from food **(1)**. The energy transfer is mechanical **(1)**. The chemical store is transferred to an increasing kinetic store due to its motion and an increasing gravitational potential energy store **(1)** due to the increase in height. Finally, all of the chemical store has been transferred to gravitational store with some being wasted as a thermal store **(1)** and dissipated to the surroundings **(1)**.

18. Efficient heat transfer

1 efficiency = (14 J ÷ 20 J) × 100% **(1)** = 0.7 or 70% **(1)**

2 a kettle transfers most of the electrical energy to thermal energy for heating the water in the kettle **(1)** but some thermal energy is always transferred to the surroundings **(1)** instead of being transferred to the water, so it must be less than 100% efficient **(1)**

19. Energy resources

1 (a) Diagram should show gpe → kinetic → electrical → (thermal energy as wasted) **(3)**

(b) Diagram should show chemical → heat or ... chemical → electrical or ... chemical → kinetic **(3)**

2 biofuels advantages: reliable, can be replaced, can be readily grown and converted to fuels for use in cars, can be stored, renewable, cheap **(any 1 for 1 mark)**; biofuels disadvantages: gives off CO_2 / greenhouse gases when burned, not renewable if not regrown **(any 1 for 1 mark)**; solar advantages: renewable, almost infinite energy supply, no cost for the solar energy **(any 1 for 1 mark)**; solar disadvantages: solar cells are expensive to make and buy, unreliable (when not sunny or at night), difficult to store the energy in electrical form, production of cells in manufacturing will pollute **(any 1 for 1 mark)**

20. Patterns of energy use

1 two of: threats to food supplies for those who need them **(1)**, global warming and floods due to using fossil fuels **(1)**, running out of non-renewable resources **(1)**

2 Wood has been readily available **(1)** for thousands of years **(1)** and has not had to be discovered, mined or have any extra technology developed (e.g. nuclear reactors/power stations) in order for it to be used to produce thermal energy **(1)**.

21. Potential and kinetic energy

1 $\Delta GPE = m \times g \times \Delta h$ = 5 kg × 10 N/kg × 18 m **(1)** = 900 **(1)** J **(1)**

2 30 km/h = 30 000 m ÷ 3600 s = $8.\dot{3}$ m/s **(1)**, KE = $\frac{1}{2}$ × 80 kg × $8.\dot{3}$ m/s **(1)** = 2778 **(1)** J **(1)**

3 KE = ΔGPE, $\frac{1}{2} \times m \times v^2 = m \times g \times \Delta h$ **(1)**, so $v^2 = 2 \times g \times \Delta h$ **(1)** 2 × 10 N/kg × 34 m **(1)**, v = $\sqrt{680}$ = 26 m/s **(1)**

22. Extended response – Conservation of energy

*Answer could include the following points (for all six marks, two advantages and two disadvantages with good supporting explanation should be supplied):

- advantages of wind – renewable, no fuel needed, no carbon dioxide gas emissions
- disadvantages – unsightly, unreliable
- advantages of coal – efficient, reliable
- disadvantages – non-renewable, carbon dioxide and sulfur dioxide gas emissions

23. Waves

1 (a) amplitude **(1)** wavelength **(1)** for correct shape **(1)**

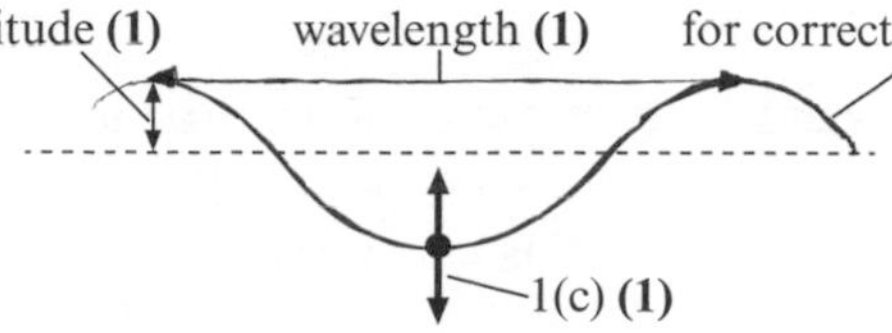

(b) ← or → **(1)** (c) arrow shown above

2 2.2 mm **(1)** *Any answer between 2.1 and 2.4 is acceptable.*

24. Wave equations

1 λ = 330 m/s ÷ 100 Hz **(1)** = 3.3 **(1)** m **(1)**

2 1 minute = 60 seconds **(1)**; x = 25 m/s × 60 s **(1)** = 1500 **(1)** m **(1)**

25. Measuring wave velocity

1 wave speed = frequency × wavelength = 4 Hz × 0.08 m **(1)** = 0.32 **(1)** m/s **(1)**

2 Speed of sound in air is found by measuring a distance for the wave to travel and a time over which the distance is covered **(1)**. Having a large distance **(1)** and a large time **(1)** for these measurements using equipment with a high degree of accuracy **(1)** will lead to a low percentage error.

26. Waves and boundaries

1 They can be reflected, refracted, transmitted or absorbed **(2 marks for all four, 1 mark for three)**.

2 Speed will change **(1)**, direction may change (unless travelling along the normal) **(1)** and wavelength of the wave will change **(1)**.

3 (a) It stops infrared radiation reaching us **(1)** which could cause overheating/death **(1)**. (b) Two of: it stops infrared radiation getting to us from space that could help us detect other objects in space **(1)**, it may be useful when it is too cold to keep people or crops warm, etc. **(1)**, it may have an impact on studies of the atmosphere or space, e.g. meteorology or climatology studies **(1)**.

27. Sound waves and the ear

1 The kinetic energy of the air molecules carrying the sound waves **(1)** is transferred into electrical energy **(1)** by vibrations inside the ear, and the electrical signal is taken to the brain **(1)**.

2 An object vibrates/moves backwards and forwards **(1)** pushing the air molecules into a series of compressions and rarefactions **(1)** that are transmitted as a pressure wave **(1)**.

3 Sound waves have the same frequency as the vibrations that cause them **(1)** and this does not change when a sound wave passes from one material to another **(1)**, whereas wave speed does **(1)** as the wave speed/wavelength is different in different materials **(1)**.

28. Ultrasound and infrasound

1 Infrasound, ultrasound and sound are all longitudinal waves. **(1)** Infrasound has lower frequencies (or longer wavelengths) than normal sound. **(1)** Ultrasound has higher frequencies (or shorter wavelengths) than normal sounds. **(1)**

2 Ultrasound waves are sent into the body from a transmitter **(1)**. Ultrasound waves are reflected at a boundary between materials of different density, e.g. blood and bone **(1)**. The reflected ultrasound is then detected by the receiver **(1)**. An image is produced on a computer monitor based on the information from the reflected ultrasound **(1)**.

3 speed = distance ÷ time = 284 m ÷ 0.2 s **(1)** = 1420 **(1)** m/s **(1)**

29. Sound wave calculations

1 Sound waves are longitudinal waves that require a medium in which particles will vibrate **(1)** to transfer the sound energy. A vacuum does not contain any particles, so these vibrations cannot occur in a vacuum **(1)**, meaning the sound wave cannot travel through it.

2 (a) Speed of wave is directly related to the wavelength of the wave, so speed in denser material = 342 m/s × (5 ÷ 2) **(1)** = 855 m/s **(1)**. (b) Assumed that there is no change in frequency of the wave **(1)** and that the wave travels faster in the material of higher density **(1)**.

30. Core practical – Waves in fluids

1 error in wavelength measured for the wave when distance being measured **(1)**, error in frequency value of the wave from generating source **(1)**, error when taking any time values with a stopwatch in order to find for the wave speed **(1)**

2 any three of: more wavelengths gives a greater distance to measure **(1)** with a measuring device of a constant accuracy/resolution **(1)** so wavelength value will be more accurate since percentage error in value obtained will be less **(1)**; example given, e.g. a metre ruler being used to measure a wavelength of 10 cm with a 1 cm scale gives a 10% error, whereas measuring 10 wavelengths would mean a 1% error (or similar) **(1)**

31. Extended response – Waves

*Answer could include the following points:

- Sound frequencies below 20 Hz are called infrasound.
- Sound frequencies above 20 000 Hz are called ultrasound.
- Infrasound can be used to explore the Earth's core and to detect volcanic eruptions.
- Ultrasound is used for medical imaging, e.g. during pre natal scans for images of a developing foetus.
- Ultrasound can be used for distance measuring in sonar equipment, e.g. depth of sea bed, shipwrecks, other obstacles or shoals of fish.
- Ultrasound is used for detecting cracks in pipes and for other quality assurance purposes.
- Any other valid example of uses of ultrasound and infrasound.

32. Reflection and refraction

1 angle of reflection = angle of incidence = 42° **(1)**

2 greater than 34° **(1)** as it speeds up and bends away from the normal **(1)**

3 two of: the ratio of the refractive indices either side of the boundary **(1)** and the optical density of the material **(1)**; material **(1)** and speed of waves in different materials **(1)**

4 (a) speeds up **(1)** when moving from more to less dense medium **(1)**, (b) decreases **(1)** when moving from less dense to more dense material **(1)**, (c) no change **(1)**

33. Total internal reflection

1 Light needs to travel from a denser material **(1)** to a less dense material **(1)** and angle of incidence needs to be above the critical angle **(1)**.

2 Light is sent into the optical fibre above the critical angle **(1)** and is totally internally reflected along the fibre to the object being viewed **(1)** where it is then reflected back up to the observer **(1)**.

3 Light rays are reflected at the interface between the fibre and the outer sheath **(1)** so that the light rays travel down into the patient and are then reflected back **(1)**. The angle at which the light waves meet this boundary must be greater than the critical angle for total internal reflection to occur **(1)** and the outer sheath must be less dense than the fibre **(1)**.

34. Colour of an object

1 Specular reflection involves the reflection of parallel rays from a smooth surface **(1)** and diffuse reflection is the reflection of rays that will not be parallel, often from a rough surface hence it is diffuse reflection from a gravel path **(1)**.

2 Red light is let through the filter (transmitted) **(1)** and all of the other colours are absorbed **(1)**.

3 You would see no colour (it would appear black) **(1)** since red would be let through the red filter but then not through the green filter **(1)**.

35. Lenses and power

1 The more curved the glass surface is **(1)**, the greater the power of the lens **(1)**.

2 As the power increases, the focal length decreases – or the reverse argument. **(1)**.

3 Light rays need to be refracted so that they are more parallel and do not need to be refracted as much by the eye **(1)**. Using a converging lens refracts the light rays so that they are parallel **(1)** and will then be focused on the retina.

36. Real and virtual images

1 a real image **(1)**

2 Light rays need to pass through the lens and be brought to a focus at a point on the other side of the lens **(1)**. When the object is closer to the lens than the focal point, the rays are refracted so that they are non-parallel and diverge **(1)**; they cannot be brought to a focus at a point so a real image cannot be formed **(1)**.

37. Electromagnetic spectrum

1 (a) radio waves **(1)**, (b) gamma-rays **(1)**

2 4×10^{-7} m: $f = c \div \lambda = 3 \times 10^{8}$ m/s $\div 4 \times 10^{-7}$ m **(1)** $= 7.5 \times 10^{14}$ Hz **(1)**; 7×10^{-7} m: $f = 3 \times 10^{8}$ m/s $\div 7 \times 10^{-7}$ m $= 4.3 \times 10^{14}$ Hz **(1)**

3 frequency $= 4 \times 10^{18} \div 60 = 6.7 \times 10^{16}$ Hz **(1)**; $\lambda = v \div f = 3 \times 10^{8}$ m/s $\div 6.7 \times 10^{16}$ Hz **(1)** $= 4.5 \times 10^{-9}$ m **(1)** (b) X-rays **(1)**

38. Core practical – Investigating refraction

1 its speed **(1)**, its direction **(1)**

2 angles of refraction are less than their corresponding angles of incidence **(1)**; angles of refraction are greater than the corresponding angles of refraction for those in the glass block **(1)**

3 (a) Frequency does not change **(1)** because the frequency is determined by the oscillating source that produces the wave and remains constant regardless of any change in speed or wavelength **(1)**. (b) Wavelength decreases **(1)** because the wave speed decreases and the frequency remains constant **(1)**.

39. Wave behaviour

1 be reflected, be refracted, be transmitted, be absorbed **(2 marks for all four, 1 mark for two three)**

2 They are refracted by the ionosphere **(1)** and reflected back to receivers on the surface of the Earth **(1)**.

40. Temperature and radiation

1 The rate at which the body absorbs radiation **(1)** is greater than the rate at which it radiates/emits radiation **(1)**.

2 The rates at which the Earth absorbs and radiates/emits thermal energy **(1)** are equal **(1)**.

3 Temperature is related to wavelength **(1)**. The wavelength of electromagnetic radiation in the visible **(1)** part of the EM spectrum determines the colour that is seen **(1)**, so changing the temperature changes the wavelength which changes the colour **(1)**.

41. Thermal energy and surfaces

1 Curves will be, in order of steepness, shiny silver **(1)**, shiny white **(1)**, shiny black **(1)**, dull or matt black **(1)**, as shiny is the worst emitter and dull black is the best emitter.

2 The rate **(1)** at which a hot body loses thermal energy **(1)** is directly related to its temperature **(1)**, so a constant, fixed temperature is needed at the start so that a fair test can be established for comparisons **(1)**.

42. Dangers and uses

1 X-rays are useful to doctors as they allow them to detect broken bones **(1)** but can damage cells **(1)**, which could cause them to mutate/lead to cancer **(1)**.

2 Gamma-rays can pass through the body and be detected by, e.g., a gamma camera, allowing a doctor to see if cancer is present **(1)**. Gamma-rays may cause cancer, but this is a risk worth taking if there is a possibility that the patient has a malignant cancer which may lead to death if not detected **(1)**.

3 Microwaves are absorbed by water molecules in the food **(1)** and their vibrations cause them to heat the food and cook it **(1)**. Infrared cooks food by heating the surface of the food only **(1)** before the heat then conducts **(1)** into the food, cooking it over a longer period of time **(1)**.

43. Changes and radiation

1 (a) Electrons move to higher energy levels **(1)** with the difference in energy between the levels equal to the energy of the electromagnetic radiation **(1)**. (b) An electron falls down from a higher to a lower energy level **(1)** with a photon emitted having with energy equal to the energy difference between the two levels **(1)**.

2 The energy difference between energy levels in the shells or orbits of electrons is much lower **(1)** compared with energy differences in the nuclei of atoms **(1)** when nuclear changes occur. The energy released in transitions between the energy levels of electrons is lower and leads to lower-energy visible light being emitted **(1)**, whereas the enormous energy changes involved during nuclear energy transitions results in much higher-energy gamma-rays being emitted **(1)**.

44. Extended response – Light and the electromagnetic spectrum

*Answer could include the following points:

- radio waves: including broadcasting, communications and satellite transmissions
- microwaves: including cooking, communications and satellite transmissions
- infrared: including cooking, thermal imaging, short-range communications, optical fibres, television remote controls and security systems
- visible light: including vision, photography and illumination
- ultraviolet: including security marking, fluorescent lamps, detecting forged bank notes and disinfecting water
- X-rays: including observing the internal structure of objects, airport security scanners and medical X-rays
- gamma-rays: including sterilising food and medical equipment, and the detection of cancer and its treatment

45. Structure of the atom

1 The diameter of the atom is around 100 000 times greater than the diameter of a nucleus **(1)**. The diameter of the nucleus on the poster will be around 60 cm $\div$ 100 000 **(1)** or a diameter of about 6×10^{-4} **(1)** cm **(1)**.

46. Atoms and isotopes

1 ${}^{11}_{5}B$ **(1)**

2 Both contain 7 protons **(1)**, both contain 7 electrons **(1)**, nitrogen-15 contains 8 neutrons, one more than nitrogen-14 **(1)**.

47. Atoms, electrons and ions

1 (a) nucleus **(1)** (b) nucleus **(1)** (c) in orbits or shells around the nucleus **(1)**

2 Atoms of an element contain the same number of protons as electrons **(1)** and have a neutral charge overall **(1)**, whereas ions of an element have the same number of protons but can gain electrons so that they have a negative charge (more electrons than protons) **(1)** or lose electrons so that they have a positive charge (more protons than electrons) **(1)**.

3 The charged particle is deflected in an electric field **(1)**, e.g. a positive ion moves towards a negative charge or plate **(1)** OR direction of motion in a magnetic field **(1)**, e.g. a positive charge will move clockwise and a negative ion anticlockwise in the same magnetic field **(1)**.

48. Ionising radiation

1 They have the greatest mass **(1)** and the greatest charge **(1)** so can remove electrons more easily from the shells of atoms **(1)**.

2 You cannot predict **(1)** when a nucleus will decay by emitting radiation **(1)**.

49. Background radiation

1 medical X-rays **(1)**

2 (a) The sector for radon would be much smaller **(1)**. (b) The sector for radon would be bigger **(1)**.

3 Radon gas emits alpha particles which are highly ionising **(1)** and cause damage inside the body **(1)** but have too small a range in air to reach the body, so will not cause harm **(1)**.

50. Measuring radioactivity

1 alpha particles are more ionising than gamma-rays **(1)** so more electrons are released per collision **(1)** so it is easier to detect from the current produced **(1)**

2 If the film behind the plastic and lead is darkened then radiation has reached it **(1)**. If the radiation behind the plastic and lead has darkened then gamma-rays have been detected **(1)** and the person is at risk as these are dangerous outside the body and can lead to cellular mutations/cancer/death **(1)**.

51. Models of the atom

1 Plum pudding model with negative plums in a positive 'dough' **(1)**. Rutherford atom model with a small positive nucleus compared to overall atomic diameter **(1)**. Bohr model with electrons orbiting a positive nucleus in definite, discrete energy levels **(1)**.

2 Positive alpha particles were fired at nucleus **(1)**. Most went through undeflected **(1)**, so most of the atom must be empty space **(1)** and there is a small nucleus as only a very few alpha particles bounced back **(1)**.

3 In the Bohr model electrons orbit the nucleus in very definite, discrete energy levels **(1)**. These energy levels are the same values for atoms of a given element **(1)**. Electronic transitions between energy levels in atoms of a gas lead to certain frequencies of electromagnetic radiation being emitted or absorbed in the spectra of light from stars **(1)**. The characteristic lines identify atoms of a particular element **(1)**.

52. Beta decay

1 (a) A neutron turns into a proton **(1)** and an electron is emitted **(1)**. (b) A proton turns into a neutron **(1)** and a positron is emitted **(1)**.

2 A beta particle is an electron that is emitted from an unstable nucleus **(1)** and an electron in a stable atom is found in shells orbiting the nucleus **(1)**.

3 ${}^{66}_{28}Ni$ **(1)** $\rightarrow {}^{66}_{29}Cu$ **(1)** + ${}^{0}_{-1}e$ **(1)**

53. Radioactive decay

1 ${}^{232}_{90}Th \rightarrow {}^{228}_{88}Ra + {}^{4}_{2}He$ **(2)**

2 Final mass number will be $A - 5$ **(1)**, nucleus loses two protons and three neutrons **(1)**; atomic number of Z does not change **(1)**, number of protons decreases by 2 from alpha decay, then increases by 2 from two beta-minus decays **(1)**.

54. Half-life

1 120 Bq to 30 Bq is one quarter of the activity or two half-lives in 4 hours **(1)**, so one half-life is half of four hours or 2 hours **(1)**.

2 After 15 minutes: $\frac{1}{2}$; after 30 minutes: $\frac{1}{4}$ **(1)**; after 45 minutes: $\frac{1}{8}$ **(1)**; after 60 minutes: $\frac{1}{16}$ **(1)**.

55. Uses of radiation

1 Any three of sterilising surgical equipment, killing cancer cells, preserving fruit, smoke alarms, thickness testing, X-ray photographs, etc. **(1 mark for each)**.

2 Contains alpha particles **(1)**, which have a short range in air **(1)**, cannot reach body **(1)**.

3 Water is injected with a gamma source **(1)**. When water reaches a crack, water will pass out of the pipe into the surrounding ground **(1)**, so a higher reading of gamma-rays is detected at the surface above the crack in the pipe **(1)**.

56. Dangers of radiation

1 two of: can cause cells to become ionised, which leads to chemical reactions taking places in cells **(1)** causing them to die **(1)** or mutate **(1)**, which can lead to cancer **(1)**

2 Ionising radiation knocks electrons from shells **(1)** making them more reactive and allowing change to DNA to occur **(1)**, whereas non-ionising radiation cannot do this as there is not enough energy to knock electrons from shells **(1)**.

3 measure radiation using a semiconductor crystal as opposed to film **(1)**, record specific exposure **(1)**, do not need to be developed so can provide information immediately **(1)**

57. Contamination and irradiation

1 (a) Irradiation is the exposure of the body to ionising radiation **(1)** from a source that is outside the body **(1)**.

(b) Contamination is exposure to ionising radiation from a source that is taken into the body **(1)** by eating, breathing in, drinking, injection or via contact ionising radiation on the surface of the skin **(1)**.

2 Alpha particles are highly ionising due to their charge and mass **(1)** and will cause damage inside the body as they have a short range **(1)** and will come into contact with cells. Outside the body they are unlikely to reach the body as they have a range of only 5 cm in air and so cannot easily reach the body to cause any ionisation in cells **(1)**.

3 Some sources of background radiation do not come into contact with the body such as cosmic rays from the Sun and medical X-rays **(1)**, but other sources such as radon gas and radioisotopes in food are taken into the body **(1)**

58. Medical uses

1 a radioisotope that is taken into the body (by injection or eating/drinking it) **(1)**. It is used to monitor biological processes in the body **(1)**

2 A single high-energy beam would have a much higher energy **(1)** so that it would destroy all cells it came into contact with when fired at the body **(1)**, whereas many low-energy beams will not have sufficient photon energy by themselves to harm healthy cells **(1)** but when they are focused at a point the energies of the individual beams add up to become a large energy that can kill the cancer tumour **(1)**.

3 Alpha particles are highly ionising so absorbtion by tissues lead to cellular mutations/cancer. Alpha particles emitted from a tracer in the body would be absorbed by the tissue around the tracer **(1)** and so could not be detected outside the body **(1)**.

59. Nuclear power

1 Energy is needed to build **(1)** and decommission **(1)** nuclear power stations, so unless all this energy comes from nuclear power stations (or renewable resources) some carbon dioxide will be added to the atmosphere from burning fossil fuels **(1)**.

2 any four from: nuclear radiation in the form of gamma-rays **(1)** can escape from nuclear power stations in small amounts **(1)** and nuclear fallout from nuclear reactors **(1)** can be produced in major nuclear disasters **(1)**, from any radioactive waste produced in the reactor **(1)** and from the decommissioning process **(1)**

60. Nuclear fission

1 Slow-moving neutrons **(1)** are absorbed by a nucleus **(1)** which becomes unstable and splits **(1)**.

2 any four from: in a chain reaction each of the neutrons produced by fission can cause another nucleus to undergo fission **(1)** so the number of atoms undergoing fission increases very rapidly **(1)** and can cause an explosion **(1)**; in a controlled chain reaction some of the neutrons are absorbed by a different material **(1)** so only one neutron from each fission can cause another atom to fission **(1)**

3 (a) so that they are easily absorbed by nuclei/do not miss the nuclei **(1)** causing the nucleus to become unstable **(1)** OR so the nucleus will become unstable **(1)** and split, releasing thermal energy **(1)**. (b) any two of: so that the extra neutrons are absorbed as only one neutron per fission needs to go on to produce further fission **(1)** as otherwise the reaction would go out of control **(1)** leading to an explosion **(1)**

61. Nuclear power stations

1 any four from: control rods absorb some of the neutrons **(1)** produced by the fission of uranium nuclei **(1)** so these neutrons cannot cause other uranium nuclei to fission **(1)**; if the control rods are pulled out of the core they will absorb fewer neutrons **(1)** so more fission reactions will happen **(1)** and more heat energy will be released **(1)**

2 Evaluation should consider any six of the following: energy released per unit mass/kg **(1)**, radioactive waste **(1)**, greenhouse gases **(1)**, availability of fuels **(1)**, threat of nuclear disaster/terrorist attack **(1)**, decommissioning costs **(1)**, reliability of fuels **(1)**, lifetime left of fuels for use in power stations **(1)**.

62. Nuclear fusion

1 Two smaller nuclei **(1)** join together to form a more stable, heavier nucleus **(1)** and release energy as thermal energy **(1)**.

2 Fusion reactions only happen at very high temperatures and pressures, **(1)** it is very difficult to produce these conditions (for any length of time) **(1)** and currently more energy is needed to make the fusion reaction happen than it releases. **(1)**

3 Fusion results in the joining of smaller nuclei **(1)** to form a nucleus with a slightly smaller mass; the left-over mass is converted to energy **(1)**. Fission results in the splitting of a larger nucleus to form more stable daughter nuclei of smaller mass **(1)**, with the small difference in mass being converted to energy **(1)**.

63. Extended response – Radioactivity

*Answer could include the following points:

- Alpha particles/alpha source must be used.
- Alpha particles ionise the air leading to the production of electrons and ions and allowing a current to flow.
- Alpha particles cannot reach the body because they are stopped by a few centimetres of air and so do not present a risk to people in the room.
- Beta particles and gamma-rays are much more penetrating and can reach the body and so would present a risk to people in the room.
- Beta and gamma are dangerous if they reach the body as they can damage cells/cause cells to mutate which could lead to illnesses such as cancer.

64. The Solar System

1 geocentric: Earth at centre, Sun and planets orbit the Earth **(1)**; heliocentric: Sun at centre, planets orbit the Sun **(1)**

2 Neptune, Uranus, Saturn, Jupiter, Mars, Earth, Venus, Mercury **(2 marks: all 8, 1 or 2 incorrect: 1 mark)**.

3 Greater distance **(1)** makes them dimmer **(1)**; recent developments in telescopes allow these to be seen **(1)**.

65. Satellites and orbits

1 Speed of a body in a circular orbit is constant **(1)** but velocity is changing as the direction of the body is constantly changing **(1)**.

2 Speed is greatest nearest the Sun **(1)** and velocity changes the most nearest the Sun **(1)** as direction changes most when nearest the Sun **(1)**. Velocity and speed are lower when further from the Sun **(1)**.

3 Orbital speed increases **(1)**, radius of orbit decreases **(1)** as centripetal force exceeds that of gravitational pull **(1)**. Orbital speed decreases **(1)**, radius of orbit increases **(1)** as gravitational pull greater than centripetal force **(1)**.

66. Theories about the Universe

1 The Universe is expanding supports both; red-shift **(1)**.

2 correct temperature **(1)** to suggest Universe started in a Big Bang **(1)** around 14 billion years ago **(1)**

3 any four from: the Universe is expanding **(1)**, so the galaxies are moving away from us **(1)** and the light from them will be red-shifted **(1)**. More distant galaxies are moving away from us faster **(1)**, so their light will have more red-shift **(1)**. The Pinwheel galaxy is further away than NGC 55, so light from it will have a greater red-shift **(1)**.

67. Doppler effect and red-shift

1 (a) siren changing its frequency **(1)** as it moves relative to the observer, frequency greater on approach to observer and lower on recession from observer **(1)**, (b) red-shift of light waves/spectral lines **(1)** for light from a galaxy or star that is moving away from us **(1)**

2 Absorption lines in the spectra of the light from the galaxy can be compared with the corresponding absorption lines of a stationary source **(1)**. The difference in the position/wavelength/frequency **(1)** can be used to calculate the speed of the galaxy **(1)**.

68. Life cycle of stars

1 (a) gravitational attraction of dust and gas **(1)**, main-sequence star/hydrogen fusion **(1)**, red giant/expansion after hydrogen burning phase **(1)**, white dwarf/collapse **(1)**, (b) same as (a) until main sequence **(2)** followed by red supergiant and supernova explosion **(1)** followed by gravitational collapse to a neutron star for a massive star and to a black hole for a supermassive star **(1)**

2 Star radiates thermal energy as it loses heat **(1)** but gains temperature due to work being done on it by gravitational collapse which raises its temperature **(1)**.

3 Gravity pulls dust and gas together to form a nebula **(1)** leading eventually to fusion **(1)** with thermal pressure balancing this when a main-sequence star due to radiation pressure from fusion **(1)** until the thermal radiation pressure exceeds the pull of gravity at the end of the star's main-sequence life and it expands **(1)**.

69. Observing the Universe

1 X-rays are absorbed by the Earth's atmosphere **(1)** so cannot be detected by Earth-based satellites **(1)**.

2 Certain wavelengths of visible light can penetrate the Earth's atmosphere **(1)** so can be detected by Earth-based telescopes **(1)**. Other wavelengths of visible light are absorbed by the atmosphere **(1)** so cannot be detected on Earth and this requires telescopes to be placed above the Earth's atmosphere in space **(1)**.

3 greater magnification – see further **(1)**; record data digitally – gather more data **(1)**; greater precision – clearer images **(1)**; detect other electromagnetic waves instead of light – can observe other objects that emit other types of EM waves **(1)**

70. Extended response – Astronomy

*Answer could include the following points.

- Low-mass stars expand to become red giants after the main sequence.
- Low-mass stars that become red giants then become white dwarfs.
- High-mass stars become red supergiants after they move off the main sequence.
- High-mass red supergiant stars undergo supernova explosions.
- Remnant cores of supernovae can become neutron stars (1.4 to 3 solar masses).
- Most massive stars become black holes (greater mass than 3 solar masses).

71. Work, energy and power

1 work = $F \times d$ = 350 N × 30 m **(1)** = 1050 **(1)** J **(1)**

2 energy = $P \times t$ = 1800 W × 8 × 60 s **(1)** = 864 000 **(1)** J **(1)**

3 power = $E \div t$ = 3600 W ÷ 60 s **(1)** = 60 **(1)** W **(1)**

4 $E = P \times t$, $\Delta E = m \times c \times \Delta T$ **(1)**, 4000 W × (46 × 60 + 40) s **(1)** = 36 kg × 4200 J/kg °C × ΔT **(1)**, ΔT = 74 **(1)** °C **(1)**

72. Extended response – Energy and forces

*Answer could include the following points:

- Refer to power as the rate of energy transfer (or similar).
- Refer to gain in gravitational potential energy as $mg\Delta h$ where Δh is the vertical distance climbed by the runner and the climber.
- Δh is a control variable.
- Time measurement needed.
- Refer to the change in gravitational potential energy for both athletes over a period of time.
- Compare their power values for running or climbing using the equation: power = energy / time or $P = mg\Delta h/t$.
- The athlete with the greatest value for $mg\Delta h/t$ (or who can transfer the most energy per second) is the most powerful.

73. Interacting forces

1 Gravity is always an attractive force, whereas magnetism and the electrostatic force can be attractive or repulsive **(1)**.

2 Friction and drag always work to oppose motion/slow down a moving object **(1)**.

3 Contact forces are any two of: drag **(1)**, normal **(1)** and upthrust **(1)**. Non-contact force is the force of gravity acting on the ship **(1)**.

74. Free-body force diagrams

1 correct length shown with correct scale **(1)**, correct angle drawn to the horizontal using a protractor **(1)**, correct horizontal value found of value close to 57.5 N **(1)**, correct vertical component shown of value close to 48.2 N **(1)**

2 force of gravity acting vertically downwards **(1)**, smaller force of drag acting upwards **(1)**, both forces labelled **(1)**

75. Resultant forces

1 (a) two forces of 20 N acting parallel and upwards **(1)**, (b) two forces of 20 N acting anti-parallel **(1)**

2 It is accelerating **(1)** downwards **(1)**.

3 correct scale drawing of the two vectors **(1 each)**, resultant correctly drawn and measured as 13 N **(1)**, direction of resultant force 67.4° to the horizontal, acting left **(1)**

76. Moments

1 Clockwise and anticlockwise moments must balance **(1)**, 160 N × 1.5 m = 200 N × d **(1)**, d = 1.2 m on the other side of the pivot **(1)**.

77. Levers and gears

1 (a) seesaw or scissors **(1)**, (b) wheelbarrow **(1)**, (c) tweezers **(1)**

2 to increase the output force from the input force such as when the car is starting to move or accelerate **(1)**, to decrease output force with respect to the input force such as when the car has reached a high speed and wants to maintain a steady speed **(1)**, to change the direction of motion such as the use of gears to make the car reverse **(1)**

3 reference to higher output force than input force/force multiplier **(1)**, reference to moments **(1)**, reference to moment = $F \times d$ **(1)**, reference to principle of moments/if clockwise and anticlockwise moments are the same, then output F is greater since output d (distance from load to the pivot) is smaller **(1)**

78. Extended response – Forces and their effects

*Answer could include the following points:

- Contact forces involve surfaces touching or being in contact.
- Non-contact forces involve bodies interacting at a distance and not being in contact.
- Examples of contact forces include friction, upthrust, reaction force.
- Examples of non-contact forces are gravitational force, magnetic force, electrostatic force.
- Contact forces such as reaction force and upthrust, if balanced, will cause a body to remain still or float.
- Contact forces, if balanced, can cause a body to move at a constant speed.
- Non-contact force of gravity will cause a body to accelerate towards Earth if there is a resultant force.

- Non-contact force of gravity will cause a body to move at a constant speed if the force of gravity is balanced by opposing frictional or drag force.
- Contact force of friction opposes the motion of a body and tries to slow it down.

Other examples of balanced/unbalanced forces used as examples for magnetism or electrostatic force.

79. Circuit symbols

1 filament lamp **(1)**, motor **(1)**, LED **(1)**

2 ammeters in series **(1)**, voltmeters in parallel **(1)**

3 circuit to contain a cell **(1)**, lamp **(1)**, LDR **(1)**

80. Series and parallel circuits

1 Series: current same at all points **(1)**, potential differences across components add to give potential difference across cell **(1)**. Parallel: current splits at a junction **(1)**, potential difference across each branch is same as potential difference across cell **(1)**.

2 Current will stop flowing in series if one bulb breaks **(1)**, so parallel arrangement so that lights stay on **(1)** if one or more bulbs break **(1)**.

3 More bulbs in parallel means a higher current **(1)** is drawn from the cell, so finite amount of stored chemical energy **(1)** decreases **(1)** more rapidly **(1)**.

81. Current and charge

1 charge: coulomb **(1)**; current: ampere **(1)**

2 $Q = I \times t = 0.25\,\text{A} \times (60 \times 60)\,\text{s}$ **(1)** = 900 **(1)** C **(1)**

3 $Q = I \times t$ so $t = Q \div I$ **(1)** $= 3 \times 10^4\,\text{C} \div 0.25\,\text{A}$ **(1)** $= 1.2 \times 10^5$ **(1)** s **(1)**

82. Energy and charge

1 $E = Q \times V = 24\,\text{C} \times 6\,\text{V}$ **(1)** = 144 **(1)** J **(1)**

2 $V = E \div Q$ **(1)**, so since unit of energy is J and charge is C **(1)**, we get 1 V = 1 J/C **(1)**.

3 using $E = QV$ and $Q = It$, $E = VIt$ **(1)**, $500\,\text{J} = 18\,\text{V} \times 240\,\text{s} \times I$ **(1)**, $I = 0.12$ **(1)** A **(1)**

83. Ohm's law

1 $V = I \times R = 3.2\,\text{A} \times 18\,\Omega$ **(1)** = 57.6 **(1)** V **(1)**

2 $R = V \div I = 28\,\text{V} \div 0.4\,\text{A}$ **(1)** = 70 **(1)** Ω **(1)**

84. Resistors

1 $I = V \div R = 12\,\text{V} \div 120\,\Omega$ **(1)** = 0.1 **(1)** A **(1)**

85. I–V graphs

1 (a) graph of S-shaped curve through origin of *I* against *V* for filament lamp **(1)**, labelled axes **(1)**. (b) As the current goes up, the temperature goes up **(1)** and as the temperature goes up the resistance goes up **(1)** and the gradient goes down **(1)**

86. Core practical – Electrical circuits

1 Current will flow through the components **(1)** in either arrangement, with the shape of the I–V graph being determined by the structure **(1)** of the component and whether it obeys Ohm's law or not **(1)**, not on how it is arranged in a circuit, since heat will still be dissipated in each circuit **(1)**.

2 As the p.d. increases across the filament lamp, the current through the filament lamp also increases **(1)** and as current increases, the energy is dissipated as heat increases leading to more vibrations of the metal ions in the filament **(1)**, which increases the resistance of the lamp **(1)** so the current increases by less and less per unit increase in p.d. **(1)** leading to a graph of decreasing gradient as the p.d. gets bigger and bigger **(1)**.

87. The LDR and the thermistor

1 (a) temperature **(1)**, (b) light intensity **(1)**

2 Provide an example of a circuit where temperature **(1)** and light intensity **(1)** vary their resistances **(1)** to turn on an output device **(1)**. The marks can only be awarded if the operation of the circuit is explained in terms of changing light levels and temperature. An example would be for a circuit that turns a light/heater on or off in a greenhouse, for example.

88. Current heating effect

1 any three devices, e.g. oven, iron, heater, hair dryer, toaster, grill **(1 mark for each)**

2 Large current **(1)**, transfers much thermal energy **(1)** which may lead to a fire **(1)**.

3 Greater current can flow **(1)**, power proportional to current squared **(1)**, more energy dissipated as heat **(1)**.

89. Energy and power

1 $I = P \div V = 3000\,\text{W} \div 230\,\text{V}$ **(1)** = 13 **(1)** A **(1)**

2 $P = E \div t = 100\,000\,\text{J} \div 180\,\text{s} = 556\,\text{W}$ **(1)**, $P = I \times V$, so $I = 556\,\text{W} \div 230\,\text{V} = 2.4\,\text{A}$ **(1)**, $R = P \div I^2 = 556\,\text{W} \div 2.4^2 = 97$ **(1)** Ω **(1)**

90. A.c. and d.c. circuits

1 (a) any three devices, e.g. remote controls, torches, mobile phones, calculators, watches, etc. **(1 mark each)**. (b) any three mains-operated devices, e.g. electric oven, microwave oven, fridge, freezer, tumble dryer, hairdryer, TV, etc. **(1 mark each)**

2 (a) They both transfer energy **(1)**. (b) Potential difference in a.c. alternates **(1)** but in d.c. it is constant **(1)**.

3 The potential difference of a.c. can be changed using transformers **(1)** so that energy losses while transmitting energy over the National Grid **(1)** are minimised **(1)**.

91. Mains electricity and the plug

1 (a) brown **(1)** (b) blue **(1)** (c) yellow and green **(1)**

2 advantage of fuse: cheaper **(1)**; advantage of circuit breaker – one of: more sensitive to current **(1)**, does not need to be replaced (reset by pressing a switch) **(1)**, more reliable **(1)**, responds faster **(1)**

3 Live wire touches metal case **(1)**, earth wire forms circuit with live wire and pulls a large current through the fuse **(1)**, large current melts fuse and isolates appliance **(1)**, device now at 0 V and no longer dangerous to user **(1)**.

92. Extended response – Electricity and circuits

*Answer could include the following points:

- Fuse will melt if the current entering an appliance is too big.
- Earth wire pulls a large current through a fuse and melts the fuse wire if the live wire comes into contact with the metal casing.
- Mention that fuse/earth wire are safety devices.
- Mention fuse is connected to live wire so that device will be at 0 V when the fuse blows.
- Fuse/earth wire isolate the appliance so that device case cannot become live and so protects the user from the electrocution / fires occurring due to overheating.
- The fuse rating should be close to, but above, the current taken by the appliance so as to avoid any risk of the device overheating/catching fire due to the current becoming too high and producing too much thermal energy.

93. Static electricity

1 (a) negatively charged **(1)**, (b) positively charged **(1)**

2 B **(1)**

3 Protons are held in the nucleus **(1)** by strong forces **(1)** so too much energy would be required **(1)** to transfer them.

4 Polythene is rubbed and becomes negatively charged **(1)** as it gains electrons; opposite charges attract **(1)**, so suspended rod will move towards the polythene **(1)** if positively charged.

94. Electrostatic phenomena

1 Plastic is an insulator and will keep any electrons that are rubbed onto it **(1)**; metal is a conductor so isolated electrons will not stay on it **(1)**.

2 Friction **(1)** between the carpet and shoes causes electrons **(1)** to move onto your shoes which are insulators **(1)**. These are earthed when a conductor is touched **(1)**. The current caused by the movement of the electrons causes the shock **(1)**.

3 Clouds become charged due to electrons being transferred between ice particles by friction **(1)** so one part of each cloud is positive and one part is negative **(1)** and a potential difference is set up between the opposite charges on each cloud **(1)** so there will be a discharge of electrons as lightning **(1)**.

95. Electrostatics uses and dangers

1 The spray droplets gain electrons when they pass through the nozzle **(1)** and these all have the same charge, causing the droplets to repel and spread out **(1)**.

2 The nozzle of the sprayer is connected to an electrical supply/ potential difference **(1)**. As droplets of insecticide pass through the nozzle they will either lose or gain electrons and so all have the same charge **(1)** causing them to repel each other **(1)** leading to a fine spray **(1)** that covers a large area of the plants.

3 Friction of plane with air meaning it can become charged **(1)**, which could ignite fuel **(1)** and cause an explosion **(1)**

96. Electric fields

1 (a) Greater mass means less acceleration **(1)** from $a = F/m$ **(1)**.

(b) It would move in the opposite direction **(1)** with the same initial acceleration **(1)**.

2 An electric field is created in the region around a charged particle **(1)**. If another charged particle is placed in (the region of) this electric field then it will experience a non-contact force **(1)** which will cause it to be attracted or repelled **(1)**. Electrons (negatively charged) are attracted to the positive charge which explains the phenomenon of static electricity **(1)**.

97. Extended response – Static electricity

*Answer could include the following points:

- Insulators can be charged if rubbed with a cloth, for example.
- Conductors cannot remain charged if rubbed.
- Electrons will move onto or off the insulator.
- Materials become negatively charged if electrons are gained or positively charged if they lose electrons.
- Refer to electrostatic induction/charge at surface of one material inducing opposite charge at the surface of the other (e.g. the tissue paper).
- Refer to attraction due to opposite charges close to each other.
- Refer to this happening for materials where induction can occur in insulators or metals by providing a suitable example.

98. Magnets and magnetic fields

1 (a) similar to the diagram for the bar magnet **(1)** but with fewer field lines **(1)**; (b) similar to the diagram for the uniform field **(1)** but with more field lines **(1)**

2 The permanent magnet always induces **(1)** the opposite pole **(1)** next to its pole causing it to attract **(1)**.

3 The permanent magnetic field of the magnets needs to interact **(1)** with the magnetic field around the coil once the current has started to flow in it **(1)**.

99. Current and magnetism

1 similar to diagram on page 99 **(1)**, but with the arrows in the opposite direction **(1)**

2 Current doubles so field doubles **(1)**, distance doubles so field halves **(1)**, so overall there is no change **(1)**.

3 Increase the size of the current flowing **(1)**, increase the number of turns of wire per metre on the coil **(1)**, insert a magnetic material as the core **(1)**.

100. Current, magnetism and force

1 A magnetic field is set up around a wire when a current flows through it **(1)** and this will interact with the field from the permanent magnet and experience a force **(1)**.

2 $F = B \times I \times L = 1.2 \times 10^{-3}\,\text{T} \times 3.6\,\text{A} \times 0.5\,\text{m}$ **(1)** $= 2.16 \times 10^{-3}$ **(1)** N **(1)**

3 (a) Reverse direction of current **(1)**, reverse direction of magnetic field **(1)**. (b) Change the size of the current **(1)**, change the size of the magnetic field strength **(1)**, change the length of the coil inside the magnetic field **(1)**.

101. Extended response – Magnetism and the motor effect

*Answer could include the following points:

- The closer a magnet is to an object, the stronger the magnetic field that it will experience, so the greater the mass it can attract.
- A greater electric current will cause the strength of the magnetic field around an electromagnet to be greater.
- More turns on the coil or a coiled wire instead of a straight wire increases the strength of the magnetic field from the electromagnetic field, so a greater mass can be picked up or attracted.
- Using a soft iron core will increase the strength of the magnetic field around a solenoid/electromagnet so more mass can be attracted.
- Mention that certain materials where magnetism can be induced will be attracted, whereas others will not – iron, cobalt, nickel, steel will be, but others will not be.
- Mention that certain materials that are themselves permanent magnets will be attracted based on their magnetic field strengths, even when no current flows through an electromagnet.

102. Electromagnetic induction

1 a magnet or magnetic field **(1)**, a conductor **(1)**, movement of the magnet relative to the conductor **(1)**

2 any three of: stronger magnet **(1)**, rotates with a constant frequency of 50 Hz **(1)**, driven by a turbine **(1)**, higher voltage output **(1)**

103. Microphones and loudspeakers

1 (a) A microphone will receive waves from 20 Hz to 20 kHz **(1)** and convert them from vibrations / sound waves to alternating electrical current or potential differences of the same frequency **(1)**. (b) A loudspeaker will convert alternating electrical signals between 20 Hz and 20 kHz **(1)** to sound waves and transfer them as vibrations of the same frequency **(1)**.

*It may be the case that microphones and loudspeakers are able to receive or transfer vibrations that are either side of this range, but these will not be audible to humans.

2 Energy transfer diagram should show: vibrations (kinetic energy store) **(1)** → mechanical transfer as sound waves **(1)** → magnetic energy store + kinetic energy store **(1)** → electrical energy **(1)**.

104. Transformers

1 $V_p = V_s \times N_p \div N_s$ **(1)** = 18 V × 40 ÷ 800 **(1)** = 0.9 V **(1)**

2 assuming 100% efficiency **(1)**, power in primary = 3000 W, so $I_p = P \div V_p$ = 3000 W ÷ 30 000 V **(1)** = 0.1 A **(1)**

3 For electromagnetic induction to occur constantly, there needs to be a constantly changing magnetic field **(1)**. This will only happen with a.c. because it is constantly changing direction **(1)** whereas d.c. does not change direction **(1)**.

105. Transmitting electricity

1 Less energy is wasted as heat **(1)** at a lower current **(1)** because the power wasted depends on I^2R **(1)**.

2 Reference to $P = I \times V$ **(1)**, high potential difference and low current **(1)**.

106. Extended response – Electromagnetic induction

*Answer could include the following points:

- Transformer has primary and secondary coil and works with a.c. input only.
- Transformer has a soft iron core which allows electromagnetic induction to induce voltage in secondary coil based on a.c. input in primary coil.
- The voltage will be stepped up/increased if the number of turns on the secondary coil is greater than on the primary coil.
- The current is stepped down when the voltage is stepped up to conserve power and not violate the principle of conservation of energy.
- Voltage is stepped up by transformers when it leaves the power station for transmission across the UK on wires.
- Voltage is stepped down to safer levels when electricity is supplied to homes, hospitals and factories.
- Current is low as power loss is minimised with a lower current: power loss is proportional to the current squared, $P \alpha I^2$.
- Lower current means thinner wires means less expensive to buy means lower cost to the consumer.

107. Changes of state

1 Thermal energy is transferred from the warmer room to the ice **(1)** and causes the ice to melt **(1)** and become 1.5 kg of liquid water **(1)**.

2 As the gas cools, the particles slow down and occupy a smaller volume **(1)** and make fewer collisions with the container they are in **(1)** so the pressure decreases too **(1)**.

3 Thermal energy is supplied to the copper metal, bonds are broken and it turns from a solid to a liquid at a constant temperature **(1)** once it reaches its melting point **(1)**. Further heating of the liquid copper causes its temperature **(1)** to rise until it reaches its boiling point, at which point it turns from a liquid to a gas at a constant temperature **(1)**. If the gaseous copper atoms continue to be heated then the temperature of the gas will increase **(1)**.

108. Density

1 density = mass ÷ volume = 862 g ÷ 765 cm^3 **(1)** = 1.1 **(1)** g/cm^3 **(1)**

2 50 000 cm^3 = 5000 ÷ 1 000 000 m^3 = 0.05 m^3 **(1)**, mass = density × volume = 0.05 m^3 × 2700 kg/m^3 **(1)** = 135 **(1)** kg **(1)**

109. Core practical – Investigating density

1 density = mass ÷ volume = 454 g ÷ 256 cm^3 **(1)** = 1.77 **(1)** g/cm^3 **(1)**

2 Mass reading can be misread by human error or by parallax error or by systematic error **(1)**; volume can be misread by parallax error or reading the top of the meniscus for water **(1)**; a value that is too high for the mass **(1)** or too low for the volume will lead to a value that is too high for the density **(1)**.

110. Energy and changes of state

1 ΔQ = 15 kg × 4200 J/kg °C × (74 °C – 18 °C) **(1)** = 3 528 000 **(1)** J **(1)**

2 $I \times V \times t = m \times c \times \Delta T$ **(1)**, c = 16 V × 2.8 A × (12 × 60) s ÷ (1.25 kg × 26 °C) **(1)** = 992 **(1)** J/kg °C **(1)**

111. Core practical – Thermal properties of water

1 Less thermal energy is lost to the surroundings **(1)** and more is used to heat the water **(1)** so the value obtained is closer to the true value required to cause a temperature change of 1 °C for that mass of liquid **(1)**; alternatively, state that losing more to the surroundings **(1)** requires more energy than necessary to be supplied for unit temperature change **(1)** of unit mass **(1)**.

2 Energy supplied **(1)** does not lead to an increase in the kinetic energy of the particles **(1)**, just in their arrangement as they change phase from a solid to a liquid **(1)**.

3 Diagram should show ice in a funnel over a beaker, with an electric heater in the ice **(1)** and a thermometer (which should be away from the heater) **(1)** and a stopwatch **(1)**.

112. Pressure and temperature

1 (a) 20 + 273 = 293 K **(1)**
(b) 300 – 273 = 27 °C **(1)**

2 –23 °C + 273 K = 250 K **(1)** and 227 °C + 273 K = 500 K **(1)**; average KE of gas particles is proportional to temperature in Kelvin, so the average KE doubles **(1)**.

3 (a) The average speed is slower **(1)**. (b) The pressure is reduced **(1)** because there are fewer collisions/the particles do not hit the walls as hard **(1)**.

113. Volume and pressure

1 $P_2 = P_1 \times V_1 \div V_2$ **(1)** 200 000 Pa × 28 cm^3 ÷ 44.8 cm^3 **(1)** = 125 000 Pa (or 125 kPa) **(1)**

2 (a) A gas compressed slowly **(1)** will change volume at a constant temperature as the thermal energy from the compression will be able to leave the system **(1)**. (b) A quick compression of a gas **(1)** means that work is done on the gas but in a tiny time period so that the increase in energy is the extra kinetic energy of the particles of gas **(1)**.

3 Assuming that the amount of gas in the bubble remains constant **(1)** and the temperature of the drink is the same throughout **(1)**, the volume increases as the bubble moves to the top of the liquid **(1)**; this is because the pressure in a liquid decreases close to the surface **(1)**.

114. Extended response – Particle model

*Answer could include the following points:

- Pressure is caused by moving particles colliding with the sides of a container.
- Pressure is force per unit area.
- Decreasing the volume of a gas quickly means that work is being done on the gas.
- When work is done on the gas the kinetic energy/speed of the particles increases.

- Faster moving particles collide with the walls/area more frequently, leading to an increase in pressure.
- Increasing the temperature of a gas at constant volume means that the particles have greater kinetic energy and speed and so collide more frequently with the area, so greater pressure.
- The internal energy of a gas is entirely kinetic.

115. Elastic and inelastic distortion

1 (a) two equal forces **(1)** of tension acting in opposite directions **(1)**, (b) two equal forces **(1)** acting towards one another **(1)**

2 (a) any two suitable examples, e.g. springs **(1)**, elastic bands **(1)**, rubber **(1)**, skin **(1)**, (b) any two suitable examples, e.g. springs beyond elastic limit **(1)**, putty **(1)**, Plasticine **(1)**, bread dough **(1)**, wet chewing gum **(1)**

3 Forces applied lead to extension of material beyond elastic limit **(1)**, which means that forces of attraction between planes of atoms or ions **(1)** are no longer strong enough for the materials to return to the original structure and so they remain permanently distorted **(1)**.

116. Springs

1 (a) $F = k \times x = 30\,\text{N/m} \times 0.2\,\text{m}$ **(1)** = 6 **(1)** N **(1)**

(b) work done $= \frac{1}{2}k \times x^2 = \frac{1}{2} \times 30\,\text{N/m} \times (0.2\,\text{m})^2$ **(1)** = 0.6 **(1)** J **(1)**

2 The greater the spring constant, the stiffer the spring **(1)** since a greater spring constant means more force is needed **(1)** to provide the same extension compared with a spring of lower spring constant **(1)**.

3 The force applied takes the spring beyond its elastic limit **(1)** and so the material will remain permanently deformed **(1)** and not return to its original shape **(1)**.

117. Core practical – Forces and springs

1 energy stored $= \frac{1}{2}kx^2 = \frac{1}{2} \times 0.8\,\text{N/m} \times (0.48\,\text{m})^2$ **(1)** = 0.09 **(1)** J **(1)**

2 Yes **(1)**, as energy is still being stored and there is a compression rather than an extension **(1)**, so the size of the compression/length used in the equations to determine the spring constant or the energy stored would be less than the original length **(1)**.

118. Upthrust and pressure

1 $P = F \div A$, so $A = F \div P$ **(1)** 600 N ÷ 20 000 Pa **(1)** = 0.03 **(1)** m^2 **(1)**

2 A has a density less than water, so will float **(1)** and displace a mass of water greater than its mass **(1)**. B has neutral buoyancy **(1)** as it displaces the same mass of water as its own mass **(1)**. C will sink as it has a greater density than water **(1)** and displaces a mass of water less than its own **(1)**.

119. Pressure and fluids

1 (a) Pressure increases as depth increase **(1)**. (b) Pressure increases as density increases **(1)**.

2 $P = h \times \rho \times g = 0.76\,\text{m} \times 13\,600\,\text{kg/m}^3 \times 10\,\text{N/kg}$ **(1)** = 103 360 **(1)** Pa **(1)**

3 $P = F \div A$ and $W = F = m \times g$, so we obtain $P = m \times g \div A$ **(1)**. Since $V = A \times h$, we obtain $P = m \times g \div (V \div h)$ **(1)**, which gives $P = m \times g \times h \div V$ **(1)**. Since $\rho = m \div V$ we get $P = h \times \rho \times g$ **(1)**

120. Extended response – Forces and matter

*Answer could include the following points:

- Distortion is when a force causes a body to extend or change shape.
- The extension is directly proportional to the load/force applied for elastic distortion.
- The body/spring/material will return to its original shape when the force/load is removed.
- Inelastic distortion is when the body is extended beyond its elastic limit.
- The body will then not return to its original shape but will remain permanently extended.
- The spring constant for a spring is found from the gradient of the force–extension graph provided that the linear region only is used.
- The spring constant may be found from $k = F \div x$.
- Energy stored is found from $E = \frac{1}{2}kx^2$ or from the area beneath the line.

Physics Equations List

In your exam, you will be provided with the following list of equations. Make sure you are clear which equations will be given to you in the exam. You will need to learn the equations that aren't on the equations list.

Equations

(final velocity)^2 – (initial velocity)^2 = 2 × acceleration × distance $v^2 - u^2 = 2 \times a \times x$
force = change in momentum ÷ time $F = \frac{(mv - mu)}{t}$
energy transferred = current × potential difference × time $E = I \times V \times t$
force on a conductor at right angles to a magnetic field carrying a current = magnetic flux density × current × length $F = B \times I \times l$
potential difference across primary coil × current in primary coil = potential difference across secondary coil × current in secondary coil $V_p \times I_p = V_s \times I_s$
$\frac{\textit{potential difference across primary coil}}{\textit{potential difference across secondary coil}} = \frac{\textit{number of turns in primary coil}}{\textit{number of turns in secondary coil}}$ $\frac{V_p}{V_s} = \frac{N_p}{N_s}$
change in thermal energy = mass × specific heat capacity × change in temperature $\Delta Q = m \times c \times \Delta\theta$
thermal energy for a change of state = mass × specific latent heat $Q = m \times L$
$P_1 V_1 = P_2 V_2$ to calculate pressure or volume for gases of fixed mass at constant temperature
energy transferred in stretching = 0.5 × spring constant × (extension)^2 $E = \frac{1}{2} \times k \times x^2$
Pressure due to a column of liquid = height of column × density of liquid × gavitational field strength $P = h \times \rho \times g$